# Inventory Optimization

**Series Editors**

Nita H. Shah, Department of Mathematics, Gujarat University, Ahmedabad, Gujarat, India

Mandeep Mittal, Department of Applied Mathematics, Amity Institute of Applied Science, Amity University, Noida, India

Leopoldo Eduardo Cárdenas-Barrón, Department of Industrial and Systems Engineering, Monterrey Institute of Technology and Higher Education, Monterrey, Mexico

Inventory management is a very tedious task faced by all the organizations in any sector of the economy. It makes decisions for policies, activities and procedures in order to make sure that the right amount of each item is held in stock at any time. Many industries suffer from indiscipline in ordering and production mismatch. Providing best policy to control such mismatch would be invaluable to them.

The primary objective of this book series is to explore various effective methods for inventory control and management using optimization techniques. The series will facilitate many potential authors to become the editors or author in this book series. The series focuses on an aspect of Operations Research which does not get the importance it deserves. Most researchers working on inventory management are publishing under different topics like decision making, computational techniques and optimization techniques, production engineering etc. The series will provide the much needed platform for them to publish and reach the correct audience.

Some of the areas that the series aims to cover are:

- Inventory optimization
- Inventory management models
- Retail inventory management
- Supply chain optimization
- Logistics management
- Reverse logistics and closed-loop supply chains
- Green supply chain
- Supply chain management
- Management and control of production and logistics systems
- Datamining techniques
- Bigdata analysis in inventory management
- Artificial intelligence
- Internet of things
- Operations and logistics management
- Production and inventory management
- Artificial intelligence and expert system
- Marketing, modelling and simulation
- Information technology

This book series will publish volumes of books which will be edited and reviewed by the reputed researcher of inventory optimization area. The beginner and experienced researchers both can publish their innovative research work in the form of edited chapters in the books of this series by getting in touch with the contact person. Practitioners and industrialist can share their real time experience bolstered with case studies. The objective is to provide a platform to the practitioners, educators, researchers and industrialist to publish their valuable work in the area of inventory optimization.

This series will be beneficial for practitioners, educators and researchers. It will also be helpful for retailers/managers for improving business functions and making more accurate and realistic decisions.

More information about this series at http://www.springer.com/series/16688

Nita H. Shah · Mandeep Mittal ·
Leopoldo Eduardo Cárdenas-Barrón
Editors

# Decision Making in Inventory Management

*Editors*
Nita H. Shah
University of Gujarat
Ahmedabad, Gujarat, India

Mandeep Mittal
Amity University
Noida, Uttar Pradesh, India

Leopoldo Eduardo Cárdenas-Barrón
Department of Industrial and Systems Engineering
Tecnológico de Monterrey
Monterrey, Mexico

ISSN 2730-9347 ISSN 2730-9355 (electronic)
Inventory Optimization
ISBN 978-981-16-1731-7 ISBN 978-981-16-1729-4 (eBook)
https://doi.org/10.1007/978-981-16-1729-4



This Springer imprint is published by the registered company Springer Nature Singapore Pte Ltd.
The registered company address is: 152 Beach Road, #21-01/04 Gateway East, Singapore 189721, Singapore

# Contents

# Editors and Contributors

## About the Editors

**Prof. Dr. Nita H. Shah** received her Ph.D. in Statistics from Gujarat University in 1994. Prof. Nita is HoD of Department of Mathematics in Gujarat University, India. She is post-doctoral visiting research fellow of University of New Brunswick, Canada. Prof. Nita's research interests include inventory modeling in supply chain, robotic modeling, Mathematical modeling of infectious diseases, image processing, Dynamical systems and its applications etc. She has completed 3-UGC sponsored projects. She has published 13 monograph, 5 textbooks, and 475+ peer-reviewed research papers. Five edited books are prepared for IGI-Global and Springer. Her papers are published in high impact Elsevier, Inderscience and Taylor and Francis Journals. By the Google scholar, the total number of citations is over 3268 and the maximum number of citations for a single paper is over 174. The H-index is 25 and i-10 index is 83 up to February 2021. She has guided 28 Ph. D. Students and 15 M.Phil. Students till now. Eight students are pursuing research for their Ph.D. degree and one post-doctoral fellow of D. S. Kothari-UGC. She has travelled in USA, Singapore, Canada, South Africa, Malaysia, and Indonesia for giving talks. She is Vice-President of Operational Research Society of India. She is council member of Indian Mathematical Society.

**Mandeep Mittal** started his career in the education industry in 2000 with Amity Group. Currently, he is working as Head and Associate Professor in the Department of Mathematics, Amity Institute of Applied Sciences, Amity University Uttar Pradesh, Noida. He earned his post-doctorate from Hanyang University, South Korea, 2016, Ph.D. (2012) from the University of Delhi, India, and postgraduation in Applied Mathematics from IIT Roorkee, India (2000). He has published more than 70 research papers in International Journals and International conferences. He authored one book with Narosa Publication on C language and edited five research books with IGI Global and Springer. He is a series editor of Inventory Optimization, Springer Singapore Pvt. Ltd. He has been awarded the Best Faculty Award by the Amity School of Engineering and Technology, New Delhi for the year 2016–2017. He guided four Ph.D. scholars, and 4 students working with him in the area of Inventory Control and Management. He also served as Dean of Students Activities at Amity School of Engineering and Technology, Delhi, for nine years, and worked as Head, Department of Mathematics in the same institute for one year. He is a member of editorial boards of Revista Investigacion Operacional, Journal of Control and Systems Engineering, and Journal of Advances in Management Sciences and Information Systems. He actively participated as a core member of organizing committees in the International conferences in India and outside India.

**Leopoldo Eduardo Cárdenas-Barrón** received the B.S. degree in industrial and systems engineering, the M.Sc. degree in manufacturing systems, the M.Sc. degree in industrial engineering, the Ph.D. in industrial engineering; all degrees from Tecnológico de Monterrey. In 1996, Leopoldo Eduardo Cárdenas-Barrón joined the Industrial and Systems Engineering Department at Tecnológico de Monterrey. At present, he is currently a Professor at Industrial and Systems Engineering Department of the School of Engineering and Sciences at Tecnológico de Monterrey, Campus Monterrey, México. He is a researcher of the research group in Optimization and Data Science. He is a member of the Mexican Research National System. He was the associate director of the Industrial and Systems Engineering programme from 1999 to 2005. Moreover, he was also the associate director of the Department of Industrial and Systems Engineering from 2005 to 2009. His research activities include inventory theory, optimization and supply chain. He has published 133 papers in international journals such as International Journal of Production Economics, Applied Mathematical Modelling, Computers and Industrial Engineering, Applied Mathematics and Computation, Mathematical and Computer Modelling, Expert Systems with Applications, OMEGA, European Journal of Operational Research, Computers and Operations Research, Applied Soft Computing, Transportation Research Part E: Logistics and Transportation Review, Computers and Mathematics with Applications, Mathematical Problems in Engineering, Production Planning and Control, Journal of the Operational Research Society, among others. He is a co-author of a book related to simulation (in Spanish).

## Contributors

**Olufemi Adetunji** Department of Industrial and Systems Engineering, University of Pretoria, Pretoria, South Africa

**Navin Ahlawat** Department of Computer Science, SRM Institute of Science and Technology, Ghaziabad, U.P., India

**Rafid B. D. Al-Rikabi** Department of Industrial and Systems Engineering, University of Pretoria, Pretoria, South Africa

**Urmila Chaudhari** Government Polytechnic Dahod, Dahod, Gujarat, India

**Neha Chauhan** SRM Institute of Science and Technology, Ghaziabad, U.P., India

**Anuja A. Gupta** Institute of Management, Nirma University, Ahmedabad, India

**Nidhi Handa** Department of Mathematics and Statistics, Gurukul Kangri Vishwavidyalaya, Haridwar, Uttarakhand, India;
Department of Mathematics and Statistics, KGC, Haridwar, India

**Mrudul Y. Jani** Department of Applied Sciences, Faculty of Engineering and Technology, Parul University, Vadodara, Gujarat, India

**Mahesh Kumar Jayaswal** Department of Mathematics and Statistics, Banasthali Vidyapith, Banasthali, Rajasthan, India

**Chetan A. Jhaveri** Institute of Management, Nirma University, Ahmedabad, India

**Chandni Katariya** Department of Mathematics and Statistics, Gurukul Kangri Vishwavidyalaya, Haridwar, Uttarakhand, India

**Poonam Mishra** Department of Mathematics, School of Technology, Pandit Deendayal Petroleum University, Raisan Gandhinagar, India

**Mandeep Mittal** Department of Mathematics, Amity Institute of Applied Sciences, Amity University Uttar Pradesh, Noida, Noida, Uttar Pradesh, India

**Ekta Patel** Department of Mathematics, Gujarat University, Ahmedabad, Gujarat, India

**Neha Punetha** Department of Mathematics and Statistics, KGC, Haridwar, India

**Kavita Rabari** Department of Mathematics, Gujarat University, Ahmedabad, Gujarat, India

**Isha Sangal** Department of Mathematics and Statistics, Banasthali Vidyapith, Banasthali, Rajasthan, India

**Makoena Sebatjane** Department of Industrial and Systems Engineering, University of Pretoria, Pretoria, South Africa

**Arash Sepehri** School of Industrial Engineering, Iran University of Science and Technology, Tehran, Iran

**Nita H. Shah** Department of Mathematics, Gujarat University, Ahmedabad, Gujarat, India

**Pratik H. Shah** Department of Mathematics, Gujarat University, Ahmedabad, Gujarat, India;
Department of Mathematics, C.U. Shah Government Polytechnic, Surendranagar, Gujarat, India

**Azharuddin Shaikh** Institute of Management, Nirma University, Ahmedabad, India

**M. Kuber Singh** Department of Mathematics, D.M. College of Science (Dhanamanjuri University), Imphal, India

**S. R. Singh** Department of Mathematics, CCS University, Meerut, India

**Anupam Swami** Department of Mathematics, Government Post Graduate College, Sambhal, U.P., India

**Isha Talati** Department of Engineering and Physical Sciences, Institute of Advanced Research, Gandhinagar, India

**Ajay Singh Yadav** Department of Mathematics, SRM Institute of Science and Technology, Ghaziabad, U.P., India

**Sarma V. S. Yadavalli** Department of Industrial and Systems Engineering, University of Pretoria, Pretoria, South Africa

# Chapter 1
# Upper-Lower Bounds for the Profit of an Inventory System Under Price-Stock Life Time Dependent Demand

**Nita H. Shah, Ekta Patel, and Kavita Rabari**

**Abstract** Price is the most important factor influencing demand rate based on marketing and economic theory. Along with price, stock display is also a major factor, as displayed stocks may induce customers to purchase more due to its visibility. Moreover, the demand for perishable products depends on its freshness. However, relatively little devotion has been paid to the influence of expiration dates despite the fact that they are an important factor in consumers' purchase decisions. As a result, we develop an inventory model for perishable products in which demand explicitly in a multivariate function of price, displayed stocks, and expiration dates. We then formulate the model by determining the optimal selling price to maximize the total profit by using classical optimization method with the necessary condition given by Kuhn-Tucker. Furthermore, we discuss the optimal decisions under two scenarios: upper bound of profit and lower bound of profit by taking holding cost as a function of upper and lower bound respectively. Finally, a numerical example is demonstrated along with sensitivity analysis to describe the impact of inventory parameters on the optimal decisions.

**Keywords** Perishable products · Price · Stock and life time dependent demand · Expiration date · Lot-sizing and classical optimization method

**MSC** 90B05

## 1.1 Introduction

Inventory management for enterprises is continuously facing challenges associated with the development, quality, design, and manufacturing of new products.

Thus the demand for new products comes and goes at a faster pace. Recently, it is observed that customers are becoming more alert and cognizant about their health as their standard of living gets better than earlier, so the demand for products with a long life cycle has drastically increased in recent years. Only an increasing

N. H. Shah (✉) · E. Patel · K. Rabari
Department of Mathematics, Gujarat University, Ahmedabad, Gujarat 380009, India



N. H. Shah et al. (eds.), *Decision Making in Inventory Management*,
Inventory Optimization, https://doi.org/10.1007/978-981-16-1729-4_1

number of products are becoming subject to loss of utility, evaporation, degradation, and devaluation because of the launch of new technology or the substitutions like fashion and seasonal goods, electronic equipment, and so on. Even products like durable furniture, high technology goods, medicines, vitamins, and cosmetics are becoming victims of perishability, so managing such perishable inventory can be very challenging. To be competitive in today's grocery industry, there is a big task for getting the right product to the right place at the right time in the right condition. Determining price and order quantity jointly is recognized for perishable products as an essential way to intensify profitability and maintain competition in the market.

It is observed that the age of perishable products has a negative impact on the demand because of the loss of consumer's confidence in the product quality. Hence, in today's market, the freshness of the product has a major effect on demand. Moreover, the expiration date is one of the major concerns to assess the freshness of a product and could significantly affect its demand. Consequently, perishable products have become more and more significant fonts of revenue in the grocery industry. Fujiwara and Perera (1993) proposed an EOQ model for perishable products in which product devalues over time by considering exponential distribution. Sarker et al. (1997) developed an inventory model for perishable products by taking the negative effect of age of the on-hand socks into consideration. Later, Hsu et al. (2006) established an inventory model by considering expiration dates for deteriorating items. Bai and Kendall (2008) studied optimal shelf space allocation for perishable products in which demand is considered to be a function of displayed stock level and freshness condition. Then, Avinadav et al. (2014) explored an inventory model for perishable products by measuring product freshness until the expiration date. Dobson et al. (2017) studied an EOQ model for perishable products with age-dependent demand in which the lower and upper bound of the cycle length and profit are analyzed. After that Chen et al. (2016) studied an inventory model for perishable products in which demand is close to zero when it approaches its expiration date. Further, Feng et al. (2017) extended Chen et al. (2016) model by adding a pricing strategy.

In practice, the demand for fresh products is influenced by the stock level, as an increase in the displayed stock level attracts more customers to purchase more. Various types of inventory models have been derived to quantify this phenomenon in studying the optimal inventory policies. Baker and Urban (1988) proposed an inventory model in which the demand rate is a polynomial function form depending on the displayed stock level. Thereafter, the first EOQ model was proposed by Urban (1992) with non-zero ending inventory with displayed stock-dependent demand. Then Urban and Baker (1997) derived an EOQ model in which demand is a deterministic and multivariate function of price, time, and level of inventory. Teng and Chang (2005) extended Urban and Baker (1997) model by scrutinizing the effect of trade credit financing along with stock level. Dye and Ouyang (2005) investigated an inventory model for perishable products under stock and price-dependent demand by considering partial backlogging. Soni and Shah (2008) formulated an inventory model in which demand is partially constant and partially dependent on stock. One step ahead, Chang et al. (2010) scrutinized an optimal replenishment

**Table 1.1** Literature survey is exhibited in Table 1.1

| Authors | Demand pattern | | Deterioration | Expiration date | Variable holding cost |
|---|---|---|---|---|---|
| | Price sensitive | Stock dependent | | | |
| Agi and Soni (2020) | ✓ | ✓ | Instantaneous | ✘ | ✘ |
| Avinadav et al. (2014) | ✓ | ✘ | ✘ | ✘ | ✘ |
| Bai and Kendall (2008) | ✘ | ✓ | Instantaneous | ✘ | ✘ |
| Baker and Urban (1988) | ✘ | ✓ | ✘ | ✘ | ✘ |
| Chang et al. (2010) | ✘ | ✓ | Non-instantaneous | ✘ | ✘ |
| Chen et al. (2016) | ✘ | ✓ | Instantaneous | ✓ | ✘ |
| Cohen (1977) | ✓ | ✘ | Instantaneous | ✓ | ✘ |
| Dobson et al. (2017) | ✘ | ✘ | Instantaneous and non-instantaneous | ✘ | ✓ |
| Dye (2007) | ✓ | ✘ | Instantaneous | ✘ | ✘ |
| Dye and Ouyang (2005) | ✓ | ✓ | Instantaneous | ✘ | ✘ |
| Feng et al. (2017) | ✓ | ✓ | Instantaneous | ✓ | ✘ |
| Fujiwara and Perera (1993) | ✘ | ✘ | Instantaneous | ✘ | ✘ |
| Hsu et al. (2006) | ✓ | ✘ | Instantaneous | ✓ | ✘ |
| Maihami and Abadi (2012) | ✓ | ✘ | Non-instantaneous | ✘ | ✘ |
| Mishra and Tripathy (2012) | ✘ | ✘ | Instantaneous | ✘ | ✘ |
| Mishra (2013) | ✘ | ✘ | Instantaneous | ✘ | ✓ |
| Papachristos and Skouri (2003) | ✓ | ✘ | Instantaneous | ✘ | ✘ |
| Sarker et al. (1997) | ✘ | ✓ | Instantaneous | ✘ | ✘ |
| Soni and Shah (2008) | ✘ | ✓ | ✘ | ✘ | ✘ |
| Teng and Chang (2005) | ✓ | ✓ | Instantaneous | ✘ | ✘ |

(continued)

**Table 1.1** (continued)

| Authors | Demand pattern | | Deterioration | Expiration date | Variable holding cost |
|---|---|---|---|---|---|
| | Price sensitive | Stock dependent | | | |
| Urban (1992) | ✘ | ✓ | ✘ | ✘ | ✘ |
| Urban and Baker (1997) | ✓ | ✓ | ✘ | ✘ | ✘ |
| Wee (1999) | ✓ | ✘ | Instantaneous | ✘ | ✘ |
| Wu et al. (2016) | ✘ | ✓ | Instantaneous | ✓ | ✘ |
| Chen et al. (2020) | ✘ | ✘ | Instantaneous | ✘ | ✘ |
| Amiri et al. (2020) | ✘ | ✘ | Instantaneous | ✘ | ✘ |
| Proposed model | ✓ | ✓ | Instantaneous | ✓ | ✓ |

policy by taking stock-dependent demand for non-instantaneous perishable products. Mishra and Tripathy (2012) exploring inventory policy for time-dependent Weibull deterioration, in this study shortages are allowed and partially backlogged. After that Mishra (2013) proposed optimal inventory policies for instantaneous perishable items with the controllable deterioration rate in which demand and holding cost are time-dependent. Wu et al. (2016) established an inventory model for fresh produce in which demand is a time-varying function of its freshness, displayed volume, and expiration date.

Selling price is also a major concern to create a repeated purchasing environment in today's competitive market scenario. In this context, Cohen (1977) proposed an inventory model for ordering and pricing decisions by considering deterministic price-dependent demand. After that Wee (1999) established an inventory model for joint pricing and order quantity decision with selling price dependent demand and partial backlogging of unsatisfied demand. Papachristos and Skouri (2003) extended the work of Wee (1999) by taking demand as continuous, convex, and decreasing in selling price. Dye (2007) addressed an inventory problem with decreasing price demand in which marginal revenue is increased. In this study, demand is not affected by product age or its freshness. More recently, Maihami and Abadi (2012) investigated an inventory model in which demand is to be a function of age and price. More recently, Agi and Soni (2020) present a deterministic model for perishable items with age, stock and price-dependent demand rate. Feng et al. (2017) scrutinized pricing and lot sizing policy for perishable items in which demand is a multivariate function of price, freshness, and displayed stocks. Chen et al. (2020) proposed an inventory model for perishable products with two self-life. Amiri et al. (2020) studied an inventory model for perishable products in a two-echelon supply chain.

The Remainder of the article is structured as follows: Sect. 1.2 defines notations and assumptions. Section 1.3 formulates the mathematical model. Section 1.4 provides numerical results. Sensitivity analysis is carried out in Sect. 1.5. Section 1.6 concludes the proposed model with future research directions.

## 1.2 Notations and Assumptions

### *1.2.1 Notations*

These are the notations that are used throughout the article (Table 1.2).

### *1.2.2 Assumptions*

Proposed inventory model is constructed on the following assumptions.

- Fresh produce has been affected by many factors such as temperature, humidity, refrigeration, time in stock among others. It seems impossible to obtain an explicit freshness of the product. However, it is well-known that fresh produce has its

**Table 1.2** Notations

| | |
|---|---|
| $\alpha$ | Scale demand, $\alpha > 0$ |
| $\beta$ | Mark up, $\beta > 0$ |
| $p$ | Selling price per unit (dollars/unit) |
| $c$ | Purchase cost per unit per dollar, $p > c$ |
| $Q$ | The order quantity |
| $\eta$ | Price elasticity, $\eta > 1$ |
| $A$ | Ordering cost per order (dollars/order) |
| $h$ | Holding cost per unit per unit time in dollars |
| $h_l$ | Holding cost for lower bound per unit per unit time in dollars |
| $h_u$ | Holding cost for upper bound per unit per unit time in dollars |
| $m$ | Expiration date (in months) |
| $T$ | Cycle time (in months) |
| $I(t)$ | The inventory level at time $t \in [0, T]$ |
| TP | Total profit in dollars |
| $\text{TP}_l$ | Total profit for lower bound in dollars |
| $\text{TP}_u$ | Total profit for upper bound in dollars |

expiration date. To make the problem easy and tractable, we may assume the maximum lifetime $f(t) = \frac{m-t}{m}, 0 < t \leq m$.

- The demand rate $R(p, I(t))$ is assumed to be a function of price, stock, and life time which is given by $R(p, I(t)) = (\alpha + \beta I(t))p^{-\eta}\frac{(m-t)}{m}$, where $\alpha$ is scale demand ($\alpha > 0$), $\beta > 0$ is mark-up, $p$ is a selling price per unit, $\eta > 1$ denotes price elasticity mark-up and m is a life time of the product.
- The inventory cycle is lower than the maximum life time of the product.
- Holding cost for lower bound is defined as $h_l = \frac{p}{m} + \frac{h}{4}$ and for upper bound holding cost is $h_u = \frac{p}{m} + \frac{h}{2}$ where $p$ is a selling price per unit $m$ is a maximum life time of the product and $h$ is a constant holding cost.
- Shortages are not allowed.
- The time horizon is infinite.

## 1.3 Mathematical Model

In Sect. 1.3, an inventory model is developed where the product loses its freshness with time. Initially, at time $t = 0$, the order quantity is $Q$, that reduced due to the effect of demand which depends upon price, stock, and life time of the product and reaches zero at time $t = T$.

The differential equation governing the inventory level at time $t$ during the interval $[0, T]$ is given by

$$\frac{dI(t)}{dt} = -(\alpha + \beta I(t))p^{-\eta}\left(\frac{m-t}{m}\right), \quad 0 \leq t \leq T \tag{1.1}$$

With the boundary condition $I(T) = 0$. Solving the differential equation in (1.1), we express the inventory level as follows

$$I(t) = -\frac{\alpha}{\beta} + \frac{\alpha e^{\frac{-1}{2}\frac{\beta p^{-\eta}t(2m-t)}{m}}}{\beta e^{\frac{-1}{2}\frac{\beta p^{-\eta}T(2m-T)}{m}}}, \quad 0 \leq t \leq T \tag{1.2}$$

Thus the order quantity could be expressed as follows:

$$Q = \frac{1}{2}\frac{T\alpha\left(p^{-2\eta}T^3\beta - 4p^{-2\eta}T^2\beta m + 4p^{-2\eta}T\beta m^2 + 2p^{-\eta}mT - 4p^{-\eta}m^2\right)}{m^2} \tag{1.3}$$

Based on the above, the profit function through the cycle consists of the following terms:

Ordering cost per cycle

$$OC = A \tag{1.4}$$

Purchase cost is given by

$$\begin{aligned} PC &= cQ \\ &= \frac{1}{2}\frac{cT\alpha\left(p^{-2\eta}T^3\beta - 4p^{-2\eta}T^2\beta m + 4p^{-2\eta}T\beta m^2 + 2p^{-\eta}mT - 4p^{-\eta}m^2\right)}{m^2} \end{aligned} \tag{1.5}$$

Holding cost during the time interval $[0,\ T]$ is given by

$$\begin{aligned} \mathrm{HC} &= h\int_0^T I(T)\mathrm{d}t \\ &= h\left(\begin{aligned} &\frac{1}{40}\frac{\alpha\beta\left(p^{-\eta}\right)^2T^5}{m^2} - \frac{1}{8}\frac{\alpha\beta\left(p^{-\eta}\right)^2T^4}{m} \\ &+\frac{1}{3}\frac{1}{\beta}\left(\alpha\left(\frac{1}{2}\frac{\beta p^{-\eta}}{m} + \frac{1}{2}\frac{\beta^2\left(p^{-\eta}\right)^2\left(-(2m-T)T+m^2\right)}{m^2}\right)T^3\right) \\ &+\frac{1}{2}\alpha\frac{\left(-\beta p^{-\eta} + \frac{\beta^2\left(p^{-\eta}\right)^2(2m-T)T}{m^2}\right)T^2}{\beta} - \frac{\alpha T}{\beta} \\ &+\frac{1}{\beta}\left(\alpha\left(1 - \frac{\beta p^{-\eta}(2m-T)T}{m} + \frac{1}{2}\frac{\beta^2\left(p^{-\eta}\right)^2(2m-T)^2T^2}{m^2}\right)T\right) \end{aligned}\right) \end{aligned} \tag{1.6}$$

Sales Revenue

$$\mathrm{SR} = pQ = \frac{pT\alpha\left(p^{-2\eta}T^3\beta - 4p^{-2\eta}T^2\beta m + 4p^{-2\eta}T\beta m^2 + 2p^{-\eta}mT - 4p^{-\eta}m^2\right)}{2m^2} \tag{1.7}$$

So, from Eqs. (1.4)–(1.7) the total profit can be calculated by following equation

$$\mathrm{TP} = \frac{1}{T}(\mathrm{SR} - \mathrm{PC} - \mathrm{HC} - \mathrm{OA})$$

$$
\mathrm{TP}=\frac{1}{T}\left(\begin{array}{l}
\frac{1}{2}\frac{1}{m^2}\left(pT\alpha\left(p^{-2\eta}T^3\beta-4p^{-2\eta}T^2\beta m+4p^{-2\eta}T\beta m^2+2p^{-\eta}Tm-4p^{-\eta}m^2\right)\right)\\
-\frac{1}{2}\frac{1}{m^2}\left(cT\alpha\left(p^{-2\eta}T^3\beta-4p^{-2\eta}T^2\beta m+2p^{-\eta}Tm-4p^{-\eta}m^2\right)\right)-A\\
-h\left(\begin{array}{l}
\frac{1}{40}\frac{\alpha\beta(p^{-\eta})^2T^5}{m^2}-\frac{1}{8}\frac{\alpha\beta(p^{-\eta})^2T^4}{m}\\
+\frac{1}{3}\frac{1}{\beta}\left(\alpha\left(\frac{1}{2}\frac{\beta p^{-\eta}}{m}+\frac{1}{2}\frac{\beta^2(p^{-\eta})^2\left(-(2m-T)T+m^2\right)}{m^2}\right)T^3\right)\\
+\frac{1}{2}\alpha\frac{\left(-\beta p^{-\eta}+\frac{\beta^2(p^{-\eta})^2(2m-T)T}{m^2}\right)T^2}{\beta}-\frac{\alpha T}{\beta}\\
+\frac{1}{\beta}\left(\alpha\left(1-\frac{\beta p^{-\eta}(2m-T)T}{m}+\frac{1}{2}\frac{\beta^2(p^{-\eta})^2(2m-T)^2T^2}{m^2}\right)T\right)
\end{array}\right)
\end{array}\right) \tag{1.8}
$$

Instead of dealing with constant holding cost, the model defines the holding in terms of expiration date and selling price, based on the holding cost, proposed inventory system can be classified into the following two categories:

- **Lower bound**

Holding cost for lower bound during the interval $[0, T]$ is given by Chen et al. (2016)

$$
\mathrm{HCl}=\left(\frac{p}{m}+\frac{h}{4}\right)\int_0^T I(t)\mathrm{d}t
$$

$$
\mathrm{HCl}=\left(\frac{p}{m}+\frac{h}{4}\right)\left(\begin{array}{l}
\frac{1}{40}\frac{\alpha\beta(p^{-\eta})^2T^5}{m^2}-\frac{1}{8}\frac{\alpha\beta(p^{-\eta})^2T^4}{m}\\
+\frac{1}{3}\frac{1}{\beta}\left(\alpha\left(\frac{1}{2}\frac{\beta p^{-\eta}}{m}+\frac{1}{2}\frac{\beta^2(p^{-\eta})^2\left(-(2m-T)T+m^2\right)}{m^2}\right)T^3\right)\\
+\frac{1}{2}\alpha\frac{\left(-\beta p^{-\eta}+\frac{\beta^2(p^{-\eta})^2(2m-T)T}{m^2}\right)T^2}{\beta}-\frac{\alpha T}{\beta}\\
+\frac{1}{\beta}\left(\alpha\left(1-\frac{\beta p^{-\eta}(2m-T)T}{m}+\frac{1}{2}\frac{\beta^2(p^{-\eta})^2(2m-T)^2T^2}{m^2}\right)T\right)
\end{array}\right) \tag{1.9}
$$

The model is analyzed the lower bound for holding cost that defines the lower range of profit function which can be calculated from Eqs. (1.4), (1.5), (1.7) and (1.9), is given by the following equation

$$
\mathrm{TPl}=\frac{1}{T}(\mathrm{SR}-\mathrm{PC}-\mathrm{HCl}-\mathrm{OA})
$$

$$\mathrm{TPl}=\frac{1}{T}\left(\begin{array}{l}\frac{1}{2}\frac{1}{m^2}\left(pT\alpha\left(p^{-2\eta}T^3\beta-4p^{-2\eta}T^2\beta m+4p^{-2\eta}T\beta m^2+2p^{-\eta}Tm-4p^{-\eta}m^2\right)\right)\\-\frac{1}{2}\frac{1}{m^2}\left(cT\alpha\left(p^{-2\eta}T^3\beta-4p^{-2\eta}T^2\beta m+2p^{-\eta}Tm-4p^{-\eta}m^2\right)\right)-A\\-\left(\frac{p}{m}+\frac{h}{4}\right)\left(\begin{array}{l}\frac{1}{40}\frac{\alpha\beta(p^{-\eta})^2T^5}{m^2}-\frac{1}{8}\frac{\alpha\beta(p^{-\eta})^2T^4}{m}\\+\frac{1}{3}\frac{1}{\beta}\left(\alpha\left(\frac{1}{2}\frac{\beta p^{-\eta}}{m}+\frac{1}{2}\frac{\beta^2(p^{-\eta})^2\left(-(2m-T)T+m^2\right)}{m^2}\right)T^3\right)\\+\frac{1}{2}\alpha\frac{\left(-\beta p^{-\eta}+\frac{\beta^2(p^{-\eta})^2(2m-T)T}{m^2}\right)T^2}{\beta}-\frac{\alpha T}{\beta}\\+\frac{1}{\beta}\left(\alpha\left(1-\frac{\beta p^{-\eta}(2m-T)T}{m}+\frac{1}{2}\frac{\beta^2(p^{-\eta})^2(2m-T)^2T^2}{m^2}\right)T\right)\end{array}\right)\end{array}\right) \tag{1.10}$$

- **Upper bound**

Holding cost for upper bound during the interval $[0, T]$ is given by is Chen et al. (2016)

$$\mathrm{HCu}=\left(\frac{p}{m}+\frac{h}{2}\right)\int\limits_0^T I(t)\mathrm{d}t$$

$$\mathrm{HCu}=\left(\frac{p}{m}+\frac{h}{2}\right)\left(\begin{array}{l}\frac{1}{40}\frac{\alpha\beta(p^{-\eta})^2T^5}{m^2}-\frac{1}{8}\frac{\alpha\beta(p^{-\eta})^2T^4}{m}\\+\frac{1}{3}\frac{1}{\beta}\left(\alpha\left(\frac{1}{2}\frac{\beta p^{-\eta}}{m}+\frac{1}{2}\frac{\beta^2(p^{-\eta})^2\left(-(2m-T)T+m^2\right)}{m^2}\right)T^3\right)\\+\frac{1}{2}\alpha\frac{\left(-\beta p^{-\eta}+\frac{\beta^2(p^{-\eta})^2(2m-T)T}{m^2}\right)T^2}{\beta}-\frac{\alpha T}{\beta}\\+\frac{1}{\beta}\left(\alpha\left(1-\frac{\beta p^{-\eta}(2m-T)T}{m}+\frac{1}{2}\frac{\beta^2(p^{-\eta})^2(2m-T)^2T^2}{m^2}\right)T\right)\end{array}\right) \tag{1.11}$$

One can replace the holding cost taken in traditional model by the holding cost given in Eq. (1.11). To achieve upper bound of the total profit which is calculated from Eqs. (1.4), (1.5), (1.7) and (1.11):

$$\text{TPu} = \frac{1}{T}(\text{SR} - \text{PC} - \text{HCu} - \text{OA})$$

$$\text{TPu} = \frac{1}{T}\left(\begin{array}{l} \frac{1}{2}\frac{1}{m^2}\left(pT\alpha\left(p^{-2\eta}T^3\beta - 4p^{-2\eta}T^2\beta m + 4p^{-2\eta}T\beta m^2 + 2p^{-\eta}Tm - 4p^{-\eta}m^2\right)\right) \\ -\frac{1}{2}\frac{1}{m^2}\left(cT\alpha\left(p^{-2\eta}T^3\beta - 4p^{-2\eta}T^2\beta m + 2p^{-\eta}Tm - 4p^{-\eta}m^2\right)\right) - A \\ -\left(\frac{p}{m}+\frac{h}{2}\right)\left(\begin{array}{l} \frac{1}{40}\frac{\alpha\beta\left(p^{-\eta}\right)^2T^5}{m^2} - \frac{1}{8}\frac{\alpha\beta\left(p^{-\eta}\right)^2T^4}{m} \\ +\frac{1}{3}\frac{1}{\beta}\left(\alpha\left(\frac{1}{2}\frac{\beta p^{-\eta}}{m} + \frac{1}{2}\frac{\beta^2\left(p^{-\eta}\right)^2\left(-(2m-T)T + m^2\right)}{m^2}\right)T^3\right) \\ +\frac{1}{2}\alpha\frac{\left(-\beta p^{-\eta} + \frac{\beta^2\left(p^{-\eta}\right)^2(2m-T)T}{m^2}\right)T^2}{\beta} - \frac{\alpha T}{\beta} \\ +\frac{1}{\beta}\left(\alpha\left(1 - \frac{\beta p^{-\eta}(2m-T)T}{m} + \frac{1}{2}\frac{\beta^2\left(p^{-\eta}\right)^2(2m-T)^2T^2}{m^2}\right)T\right) \end{array}\right) \end{array}\right) \quad (1.12)$$

### 1.3.1 Optimal Solution

The model uses classical optimization method to maximize the total profit

**Step** 1: Differentiate all the three profit functions derived in Eqs. (1.8), (1.10) and (1.12) with respect to inventory parameters $T$ and $p$ partially.

**Step** 2: Solve the equations for $T$ and $p$.

**Step** 3: Allocate the values to all the inventory parameters except decision variables.

**Step** 4: Substitute in all the profit functions.

## 1.4 Numerical Validation

This section validates the proposed model with a numerical example and managerial insights are also given.

$$A = \$140, \alpha = 650, \beta = 3.5, \eta = 1.002,$$
$$m = 8\,\text{months}, h = \$5/\text{unit}, c = \$20/\text{unit}$$

In such condition the solution: cycle time $T = 6.85$ months, selling price $p = \$27.65$ / unit and total profit is $TP = \$290.59$.

Graphical representation in all the three cases: lower bound of total profit, total profit, and upper bound of total profit are validated in maple 18 as shown below:

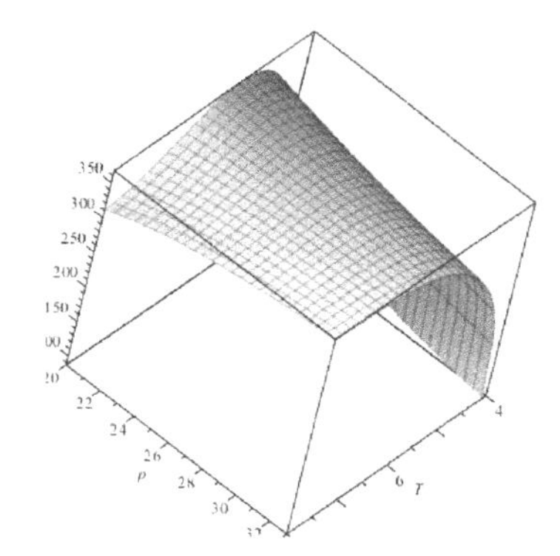


(a) Concavity of total profit w.r.to cycle time and selling price

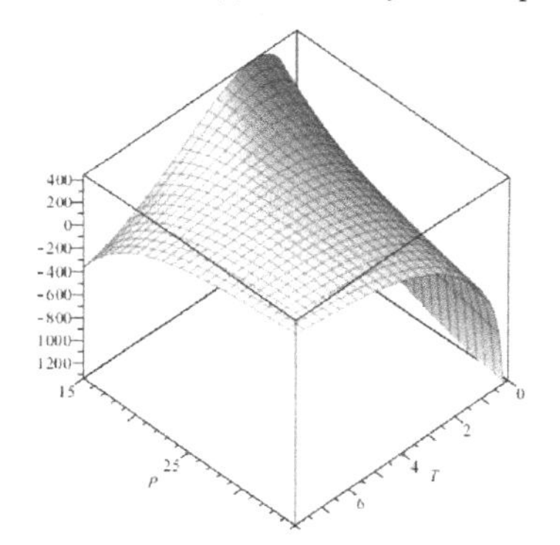


(b) Concavity of lower bound of total profit w.r.to cycle time and selling price

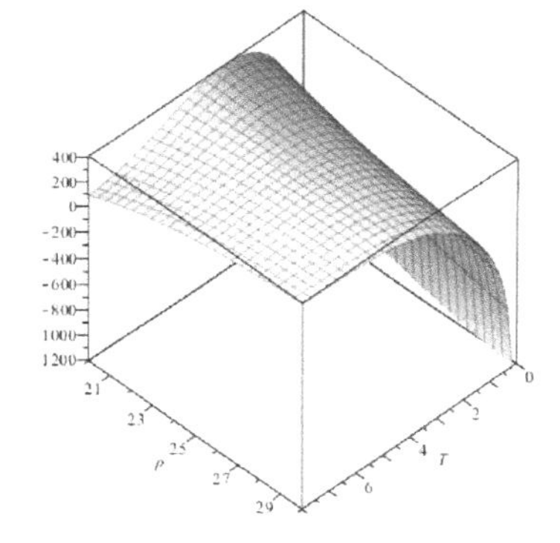


(c) Concavity of upper bound of total profit w.r.to cycle time and selling price

**Fig. 1.1** Concavity of total profit

Figure 1.1 show the concavity of total profit with respect to cycle time and selling price. Figure 1.1a is of total profit with constant holding cost. Lower bound of total profit is presented in Fig. 1.1b. Upper bound of profit is displayed in Fig. 1.1c.

## 1.5 Sensitivity Analysis

Based on the result, we performed the sensitivity analysis by changing the value of one parameter at a time by a factor of negative and positive of 10 and 20%. Effects of such changes in each parameter on the optimal solutions are studied. Based on the holding cost this analysis is categorized into two different cases:

**Case: 1 Lower bound**

The variation in cycle time, selling price, and total profit are presented in Fig. 1.2a–c respectively. It is observed from Fig. 1.2a that cycle time is more sensitive to purchase cost $c$, expiration date $m$, price elasticity $\eta$ and mark-up $\beta$. As purchase cost $c$, expiration date $m$ and price elasticity $\eta$ increases, cycle time will increase. Inventory parameters scale demand $\alpha$, ordering cost $A$ and holding cost $h$ have reasonable effects on cycle time.

Figure 1.2b, c show that with a rise in scale demand, mark-up and purchase cost, selling price as well as total profit will increase. So it is advisable for a profitable business. Expiration dates have a huge impact on the model. If the duration of the expiration date is short, a business faces financial loss in terms of the reduced profit function. Inventory parameters ordering cost $A$ and price elasticity $\eta$ plays a negative impact on profitability. An increase in ordering cost and price elasticity reduces the total profit. Hence, an increase is not preferable.

**Case: 2 Upper bound**

Figure 1.3a–c shows the change in cycle time, selling price, and total profit with respect to other inventory parameters. Figure 1.3a shows inventory parameters scale demand $\alpha$, ordering cost $A$, and holding cost $h$ have a significant effect on cycle time. An increase in purchase cost $c$, price elasticity $\eta$ and expiration date $m$ increases cycle

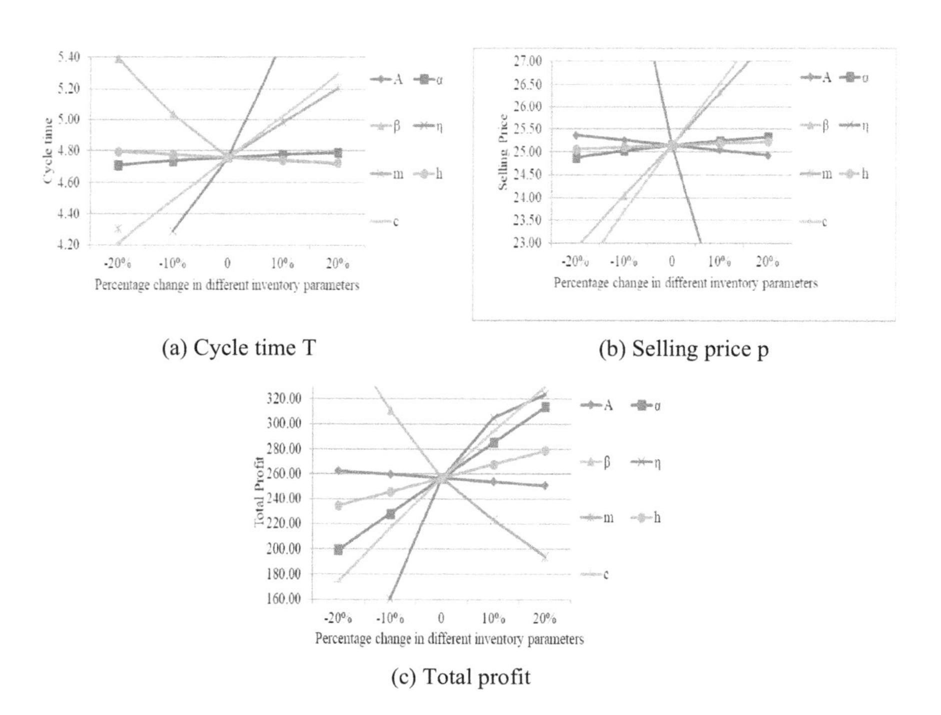

(a) Cycle time T

(b) Selling price p

(c) Total profit

**Fig. 1.2** Sensitivity analysis of lower bound

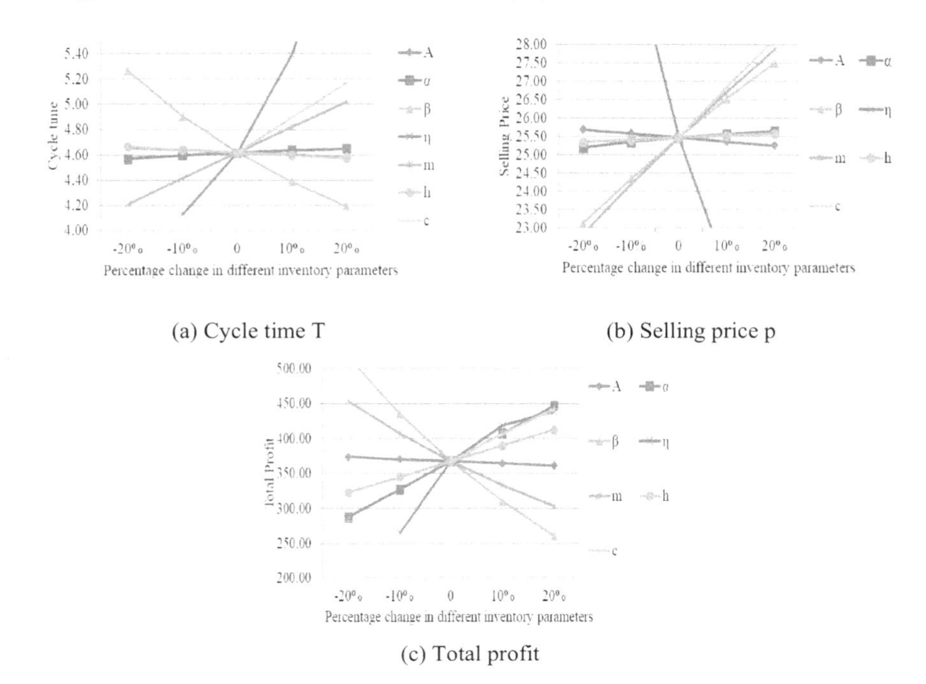


(a) Cycle time T

(b) Selling price p

(c) Total profit

**Fig. 1.3** Sensitivity analysis of upper bound

time. On the other hand, cycle time decreases with an increase of mark up $\beta$. From Fig. 1.3b, it can be shown that purchase cost $c$, expiration date $m$ and mark-up $\beta$ have a positive impact on selling price whereas price elasticity $\eta$ has a reversible effect on selling price. The total profit gets increased with the rise of scale demand $\alpha$, ordering cost $A$, holding cost $h$, purchase cost $c$ and price elasticity $\eta$. Profit will decrease for expiration date $m$ and mark up $\beta$. The rest of the parameters have a reasonable effect on total profit depicted in Fig. 1.3c.

Sensitivity analysis of cycle time, selling price, and total profit with lower and upper bound is exposed in the following figures.

From Fig. 1.4a, it is observed that total profit is not affected by increasing ordering cost in both the cases: lower bound and upper bound. In Fig. 1.4b, there is no significant change in lower bound of profit due to a change in holding cost. Figure 1.4c, d represent the effect of change in scale demand and mark-up on total profit. In both scenarios, the total profit increase with an increase in scale demand while total profit gets decreases with increases in inventory parameter mark-up.

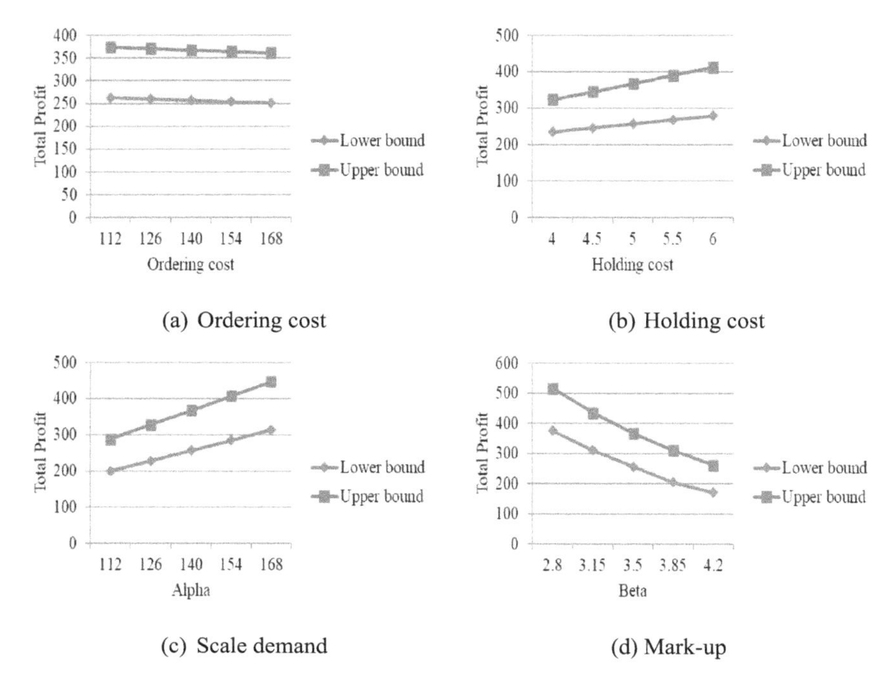


(a) Ordering cost (b) Holding cost

(c) Scale demand (d) Mark-up

**Fig. 1.4** Lower and upper bound of inventory parameters

## 1.6 Conclusion

In the model proposed here, we have explicitly taken the demand for a perishable product as a multivariate function of its selling price, stock level, and expiration date by incorporating the following facts: selling price is an important strategy for changing the customers purchasing decision, a large quantity of displayed stock motivates more sales and perishable product have a short life and cannot be sold after its expiration date. Hence, managing perishable products is a key success factor for any business to be successful. The model is analyzed analytically and graphically by maximizing the total profit. Additionally, we have represented the lower and upper bound of total profit by defining the function form of holding cost. A numerical example is given to demonstrate the applicability of the model. We have performed a sensitivity analysis to examine how each inventory parameter affects the total profit, selling price, and cycle time. From this study, it can be observed that putting so much stock on display has its own shortcoming such as loss due to holding cost and expiration date. This model can further be extended by taking more realistic assumptions like probabilistic demand rate and to strengthen the applicability of the proposed model one can add advertising strategy and quantity discount or one can expand this single-player local optimal solution to an integrated cooperative solution for two players in the supply chain.

**Acknowledgements** The authors thank DST-FIST file # MSI-097 for the technical support to the department. Second author (Ekta Patel) would like to extend sincere thanks to the Education Department, Gujarat State for providing scholarship under scheme of developing high quality research (Student Ref No: 201901380184). Third author (Kavita Rabari) is funded by a Junior Research Fellowship from the Council of Scientific & Industrial Research (file No.-09/070(0067)/2019-EMR-I).

## References

Agi MA, Soni HN (2020) Joint pricing and inventory decisions for perishable products with age-, stock-, and price-dependent demand rate. J Oper Res Soc 71(1):85–99

Amiri SAHS, Zahedi A, Kazemi M, Soroor J, Hajiaghaei-Keshteli M (2020) Determination of the optimal sales level of perishable goods in a two-echelon supply chain network. Comput Indus Eng 139:106156

Avinadav T, Herbon A, Spiegel U (2014) Optimal ordering and pricing policy for demand functions that are separable into price and inventory age. Int J Prod Econ 155:406–417

Bai R, Kendall G (2008) A model for fresh produce shelf-space allocation and inventory management with freshness-condition-dependent demand. INFORMS J Comput 20(1):78–85

Baker RA, Urban TL (1988) A deterministic inventory system with an inventory-level-dependent demand rate. J Oper Res Soc 39(9):823–831

Chang CT, Teng JT, Goyal SK (2010) Optimal replenishment policies for non-instantaneous deteriorating items with stock-dependent demand. Int J Prod Econ 123(1):62–68

Chen SC, Min J, Teng JT, Li F (2016) Inventory and shelf-space optimization for fresh produce with expiration date under freshness-and-stock-dependent demand rate. J Oper Res Soc 67(6):884–896

Chen K, Xiao T, Wang S, Lei D (2020) Inventory strategies for perishable products with two-period shelf-life and lost sales. Int J Prod Res 1–20

Cohen MA (1977) Joint pricing and ordering policy for exponentially decaying inventory with known demand. Naval Res Logistics Q 24(2):257–268

Dobson G, Pinker EJ, Yildiz O (2017) An EOQ model for perishable goods with age-dependent demand rate. Eur J Oper Res 257(1):84–88

Dye CY (2007) Joint pricing and ordering policy for a deteriorating inventory with partial backlogging. Omega 35(2):184–189

Dye CY, Ouyang LY (2005) An EOQ model for perishable items under stock-dependent selling rate and time-dependent partial backlogging. Eur J Oper Res 163(3):776–783

Feng L, Chan YL, Cárdenas-Barrón LE (2017) Pricing and lot-sizing polices for perishable goods when the demand depends on selling price, displayed stocks, and expiration date. Int J Prod Econ 185:11–20

Fujiwara O, Perera ULJSR (1993) EOQ models for continuously deteriorating products using linear and exponential penalty costs. Eur J Oper Res 70(1):104–114

Hsu PH, Wee HM, Teng HM (2006) Optimal lot sizing for deteriorating items with expiration date. J Inf Optim Sci 27(2):271–286

Maihami R, Abadi INK (2012) Joint control of inventory and its pricing for non-instantaneously deteriorating items under permissible delay in payments and partial backlogging. Math Comput Model 55(5–6):1722–1733

Mishra VK (2013) An inventory model of instantaneous deteriorating items with controllable deterioration rate for time dependent demand and holding cost. J Indus Eng Manag (JIEM) 6(2):496–506

Mishra U, Tripathy CK (2012) An inventory model for time dependent Weibull deterioration with partial backlogging. Am J Oper Res 2(2):11–15

Papachristos S, Skouri K (2003) An inventory model with deteriorating items, quantity discount, pricing and time-dependent partial backlogging. Int J Prod Econ 83(3):247–256
Sarker BR, Mukherjee S, Balan CV (1997) An order-level lot size inventory model with inventory-level dependent demand and deterioration. Int J Prod Econ 48(3):227–236
Soni H, Shah NH (2008) Optimal ordering policy for stock-dependent demand under progressive payment scheme. Eur J Oper Res 184(1):91–100
Teng JT, Chang CT (2005) Economic production quantity models for deteriorating items with price-and stock-dependent demand. Comput Oper Res 32(2):297–308
Urban TL (1992) An inventory model with an inventory-level-dependent demand rate and relaxed terminal conditions. J Oper Res Soc 43(7):721–724
Urban TL, Baker RC (1997) Optimal ordering and pricing policies in a single-period environment with multivariate demand and markdowns. Eur J Oper Res 103(3):573–583
Wee HM (1999) Deteriorating inventory model with quantity discount, pricing and partial backordering. Int J Prod Econ 59(1–3):511–518
Wu J, Chang CT, Cheng MC, Teng JT, Al-khateeb FB (2016) Inventory management for fresh produce when the time-varying demand depends on product freshness, stock level and expiration date. Int J Syst Sci Oper Logistics 3(3):138–147

# Chapter 2
# An Inventory Model for Stock and Time-Dependent Demand with Cash Discount Policy Under Learning Effect and Partial Backlogging

**Nidhi Handa, S. R. Singh, and Chandni Katariya**

**Abstract** This study considers an inventory model with stock and time-dependent demand. Stock level always plays a very vital role and affects the demand rate. Vendors usually offer different schemes to attract more customers. In this paper, the scheme of cash discount is working as promotional tool for increasing demand rate. Shortages are allowed with partial backlogging and backlogging rate present in the model is assumed as a waiting time-dependent function. To make the study more realistic learning effect is applied on holding cost. Three cases for the allowed trade credit period are described in the present paper. To illustrate the model numerical example for different cases have been discussed by using Mathematica 11.3. sensitivity analysis with respect to distinct parameters is carried out for the feasibility and the applicability of the model.

**Keywords** Inventory model · Learning effect · Deterioration · Stock and time-dependent demand · Cash discount · Trade credit · Partial backlogging · Shortages

## 2.1 Introduction

The role of demand is very vital while developing an inventory model, Available stock and time are the factors that always influence the demand. Khurana and Chaudhary (2016) proposed an inventory model using stock and price-dependent demand for deteriorating items under shortage backordering. Giri et al. (2017) introduced a vendor–buyer supply chain inventory model using time-dependent demand under preservation technology. Khurana and Chaudhary (2018) developed a deteriorating inventory model for stock and time-dependent with partial backlogging. Bardhan

N. Handa · C. Katariya (✉)
Department of Mathematics and Statistics, Gurukul Kangri Vishwavidyalaya, Haridwar, Uttarakhand 249404, India

S. R. Singh
Department of Mathematics, CCS University, Uttar Pradesh, Meerut 250001, India



N. H. Shah et al. (eds.), *Decision Making in Inventory Management*, Inventory Optimization, https://doi.org/10.1007/978-981-16-1729-4_2

et al. (2019) introduced a non-instantaneous deteriorating inventory model for stock-dependent demand under preservation technology. Handa et al. (2020) worked on an EOQ model with stock-dependent demand for trade credit policy under shortages.

Deterioration is another important factor whose part in the construction of an inventory model is very useful. Deterioration can be defined as the reduction, or spoilage in the original value of the product. Skouri et al. (2009) developed some inventory policies under Weibull deterioration rate for ramp type demand. Chowdhury et al. (2014) formulated an inventory model for price and stock-dependent demand. Mahapatra et al. (2017) introduced a model using deteriorating items based on reliability-dependent demand under partial backlogging. Rastogi et al. (2018) developed an inventory policy for non-instantaneous deteriorating items using price-sensitive demand with partial backordering.

In the construction of an inventory model, the basic assumption is that when the stock out situation occurs then the shortages that take place are either completely lost or completely backlogged which is not realistic. At the arrival of the stock some customers are interested to come back, which is known as partial backlogging. Roy and Chaudhuri (2011) studied an inventory model using price-dependent demand, Weibull deterioration, and partial backlogging. Kumar and Singh (2014) presented a two-warehouse inventory model in which demand depends upon stock level under partial backordering. Geetha and Udayakumar (2016) formulated inventory policies for non-instantaneous deteriorating products under multivariate demand rate and partial backorder. Khanna et al. (2017) proposed an inventory model using selling price-dependent demand for imperfect items under shortage backordering and trade credit. Kumar et al. (2020) studied the effect of preservation and learning on partial backordering inventory model for deteriorating items with the environment of the Covid-19 pandemic.

In today's competitive market the trade credit period offered by the seller has become a very useful incentive policy for attracting new customers. Singh et al. (2016) proposed an EOQ model allowing stock-dependent demand under trade credit policy. Shaikh (2017) introduced a deteriorating inventory model based on advertisement and price-dependent demand using partial backlogging and mixed type of trade credit. Tripathi et al. (2018) studied an inventory model for time-varying holding cost with stock dependent demand having different. Shaikh et al. (2019) introduce a Weibull distributed deteriorating inventory model allowing multivariate demand rate and trade credit period.

Learning is a realistic phenomenon that occurs naturally. Generally, it is seen that when workers accomplish the same procedure repeatedly then they learn how to performs more efficiently such phenomenon is called learning effect. Singh et al. (2013) presented an inventory model for imperfect products under the effect of inflation and learning. Singh and Rathore (2016) formulated a reverse logistic inventory model with preservation and inflation under learning effect. Goyal et al. (2017) proposed an EOQ model using advertisement-based demand under learning effect and partial backorder. Singh et al. (2020) introduced a reverse logistic inventory model for variable production under learning effect.

This paper represents an inventory model considering stock and time-dependent demand, cash discount, and partial backlogging. To make the study more realistic learning effect is applied on holding cost. Different cases for the allowed trade credit period are also described in the model. To improve the efficiency of the model numerical example for different cases and sensitivity analysis for distinct value of parameters have been discussed.

## 2.2 Assumptions

1. Demand used in the model is a function of stock and time i.e. $(\delta+\beta t+\gamma E_1(t))$.
2. Items used in the model are of decaying nature.
3. No replacement policy is allowed for deteriorating products in whole cycle period.
4. Shortages are considered with partial backlogging.
5. Deteriorating rate is constant.
6. Backlogging rate present in the model is assumed as a waiting time-dependent function.
7. This model incorporates the effect of learning on holding cost.
8. Trade credit period is allowed in the model.

## 2.3 Notations

Notations used in the model.

| | |
|---|---|
| $E(t)$ | level of inventory at any time $t$ |
| $\delta, \beta, \gamma$ | coefficients of demand |
| $Q_1$ | initial stock level |
| $Q_2$ | backorder quantity during stock out |
| $k$ | rate of deterioration |
| $\phi(\eta)$ | rate of backlogging |
| $\eta$ | waiting time up to next arrival lot |
| $T$ | cycle time |
| $u_1$ | time at which level of inventory becomes zero |
| $h_f+\frac{h_g}{n^\lambda}$ | per unit holding cost under learning effect where $\lambda>0$ |
| $s_r$ | shortage cost per unit |
| $d$ | per unit deterioration cost |
| $l_r$ | per unit lost sale cost |
| $c$ | purchasing cost per unit |
| $A$ | per order ordering cost |
| $p$ | selling price per unit |
| $U.T.P_x.$ | unit time profit |

| | |
|---|---|
| $M$ | allowed trade credit period |
| $I_c$ | rate of interest charged |
| $I_e$ | rate of interest earned |
| y | rate of cash discount. |

## 2.4 Mathematical Modelling

Figure 2.1 represents the behavior of inventory system with respect to time. $Q_1$ denotes the initial inventory level at $t = 0$. Level of inventory depletes in the interval $[0, u_1]$ for the reason of deterioration and demand. At $t = u_1$, inventory level turns into zero, and after that shortages occur with partial backlogging. The depletion of the inventory is shown in Fig. 2.1

Differential equations of the inventory system can be represented as follows

$$\frac{\mathrm{d}E_1}{\mathrm{d}t} + kE_1 = -(\delta + \beta t + \gamma E_1(t)) \quad 0 \le t \le u_1 \tag{2.1}$$

$$\frac{\mathrm{d}E_2}{\mathrm{d}t} = -(\delta + \beta t) \quad u_1 \le t \le T \tag{2.2}$$

Boundary equations are given as follows:

$$E_1(u_1) = E_2(u_1) = 0$$

Solution of Eqs. (2.1) and (2.2) are given by

$$\begin{aligned} E_1(t) = {} & \left[\delta(u_1 - t) + \frac{\beta}{2}\left(u_1^2 - t^2\right) + (k+\gamma)\left\{\frac{\delta}{2}\left(u_1^2 - t^2\right)\right.\right. \\ & \left.\left. + \frac{\beta}{2}\left(u_1^3 - t^3\right)\right\}\right]\mathrm{e}^{-(k+\gamma)t} \quad 0 \le t \le u_1 \end{aligned} \tag{2.3}$$

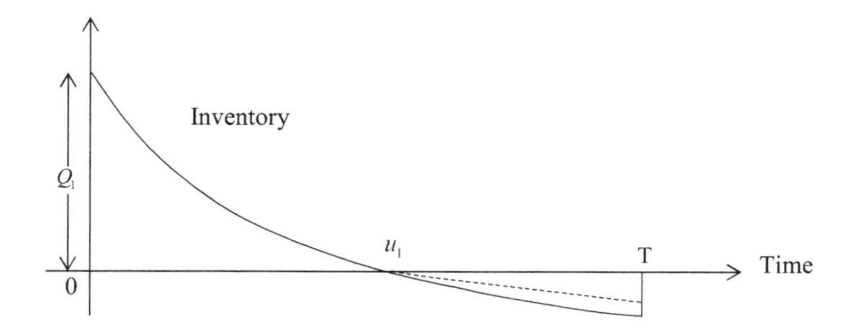


**Fig. 2.1** Inventory time graph of system

$$E_2(t) = \left[\delta(u_1 - t) + \frac{\beta}{2}\left(u_1^2 - t^2\right)\right] \quad u_1 \le t \le T \tag{2.4}$$

## 2.5 Associated Costs

**Ordering Cost**:

Ordering cost per order for the system is taken as follows:

$$O.C_x. = A \tag{2.5}$$

**Purchasing Cost**:

$Q_1$ denotes the initial inventory level at $t = 0$ and $Q_2$ for the duration $[u_1, T]$.

$$E_1(0) = Q_1 = \left\{\delta u_1 + \beta\frac{u_1^2}{2} + (k+\gamma)\left(\delta\frac{u_1^2}{2} + \beta\frac{u_1^2}{3}\right)\right\} \tag{2.6}$$

$$Q_2 = \int\limits_{u_1}^{T} (\delta + \beta t)\phi(\eta)\mathrm{d}t \tag{2.7}$$

$$= \left\{\frac{\delta}{2}\left(T^2 - u_1^2\right) + \frac{\beta}{3}\left(T^3 - u_1^3\right)\right\} \tag{2.8}$$

$$P.C_x. = \{Q_1 + Q_2\}c \tag{2.9}$$

Hence, the purchasing cost of the system is given by

$$P.C_x = \left\{\delta u_1 + \beta\frac{u_1^2}{2} + (k+\gamma)\left(\delta\frac{u_1^2}{2} + \beta\frac{u_1^2}{3}\right) + \frac{\delta}{2}\left(T^2 - u_1^2\right) + \frac{\beta}{3}\left(T^3 - u_1^3\right)\right\}c \tag{2.10}$$

**Sales Revenue**:

Sales revenue can be taken as follows:

$$S.R_x. = (Q_1 + Q_2)p \tag{2.11}$$

$$S.R_x. = \left\{\delta u_1 + \beta\frac{u_1^2}{2} + (k+\gamma)\left(\delta\frac{u_1^2}{2} + \beta\frac{u_1^2}{3}\right) + \frac{\delta}{2}\left(T^2 - u_1^2\right) + \frac{\beta}{3}\left(T^3 - u_1^3\right)\right\}p \tag{2.12}$$

**Holding Cost**:

Holding cost is considered in the duration when the system holds the inventory. Holding cost is taken as follows:

$$H.C_x. = \left(h_f + \frac{h_g}{n^\lambda}\right)\int_0^{u_1} E_1(t)\mathrm{d}t \tag{2.13}$$

$$H.C_x. = \left(h_f + \frac{h_g}{n^\lambda}\right)\left\{\delta\frac{u_1^2}{2} + \beta\frac{u_1^3}{3} + (k+\gamma)\left(\delta\frac{u_1^3}{6} + \beta\frac{u_1^4}{8}\right)\right\} \tag{2.14}$$

**Shortage Cost**:

In the inventory system shortages occur during the stock out condition when goods are not available to fulfil the customers demand. Shortage cost of the system is taken as follows:

$$S.C_x. = s_r\int_{u_1}^{T} (\delta + \beta t)\mathrm{d}t \tag{2.15}$$

$$S.C_x. = \left\{\delta(T - u_1) + \frac{\beta}{2}\left(T^2 - u_1^2\right)\right\}s_r \tag{2.16}$$

**Lost Sale Cost**:

In the inventory system lost sale cost is considered during the stock out condition when some customers fulfil their demand from other places. Lost sale cost is taken as follows:

$$L.S.C_x. = l_r\int_{u_1}^{T} (\delta + \beta t)(1 - \phi(\eta))\mathrm{d}t \tag{2.17}$$

$$L.S.C_x. = l_r\left\{\delta\frac{T^2}{2} + \beta\frac{T^3}{6} - \delta u_1 T - \beta T\frac{u_1^2}{2} + \delta\frac{u_1^2}{2} + \beta\frac{u_1^3}{3}\right\} \tag{2.18}$$

**Deterioration Cost**:

Deterioration cost is considered for those products that are deteriorated or decayed in the system. The deterioration cost is taken as follows:

$$D.C_x. = d\left\{E_1(0) - \int_0^{u_1} (\delta + \beta t)\mathrm{d}t\right\} \tag{2.19}$$

$$D.C_x. = d(k+\gamma)\left(\delta\frac{u_1^2}{2} + \beta\frac{u_1^3}{3}\right) \tag{2.20}$$

## 2.6 Permissible Delay

Trade credit period is the useful incentive policy for attracting more customers. In this time period vendor allows a certain time limit to retailer to pay all his dues. If the retailer pays all his dues before the credit limit then there will be no interest charged otherwise interest will be charged on unpaid amount. Retailer can also earn interest on sales revenue.

Two cases for allowed trade credit period are given as follows:

**Case 1**: When $M \geq u_1$ (Fig. 2.2).

For this case vendor has enough amount to settle all his payments since the credit limit period is more than the period of sold out all the stock. In this case, interest charged would be zero and interest earned in the duration [0, M] is given as follows.

$$I.V_1. = pI_e \int_0^{u_1} (\delta + \beta t + \gamma E(t))\mathrm{d}t + (M - u_1) \int_0^{u_1} (\delta + \beta t + \gamma E(t))\mathrm{d}t \tag{2.21}$$

$$\begin{aligned} pI_e & \left\{ \frac{\delta u_1^2}{2} + \frac{\beta u_1^3}{3} - \gamma\left(\frac{\delta(k+\gamma)+\beta}{8}u_1^4 + \frac{\delta u_1^3}{6} + \frac{\beta(k+\gamma)}{10}u_1^5 \right.\right. \\ & - \frac{\delta(k+\gamma)}{12}u_1^4 - \frac{(\delta(k+\gamma)^2 + \beta(k+\gamma))}{15}u_1^5 \\ & \left.\left. - \frac{\beta(k+\gamma)^2}{9}u_1^6\right)\right\} + \{(M - u_1) \\ & \left(\delta u_1 + \frac{\beta u_1^2}{2} - \gamma\left(\frac{\delta u_1^2}{2} + \frac{\delta(k+\gamma)+\beta}{3}u_1^3 - \delta(k+\gamma)\frac{u_1^3}{6}\right.\right. \\ & \left.\left.\left. - \frac{(\delta(k+\gamma)^2 + \beta(k+\gamma))}{8}u_1^4 - \frac{\beta(k+\gamma)^2}{10}u_1^5\right)\right)\right\} \end{aligned} \tag{2.22}$$

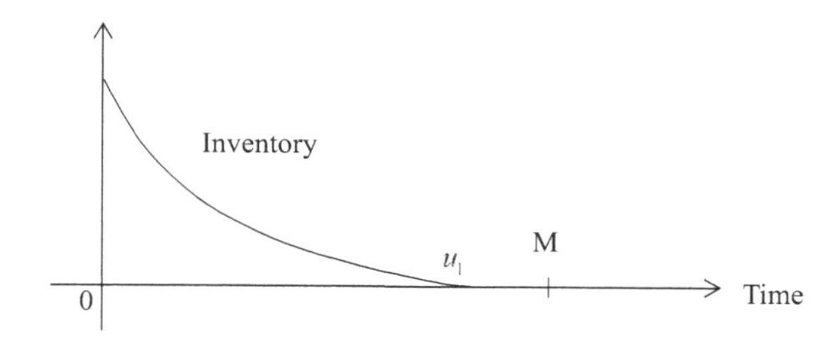


**Fig. 2.2** Inventory time graph when $(M \geq u_1)$

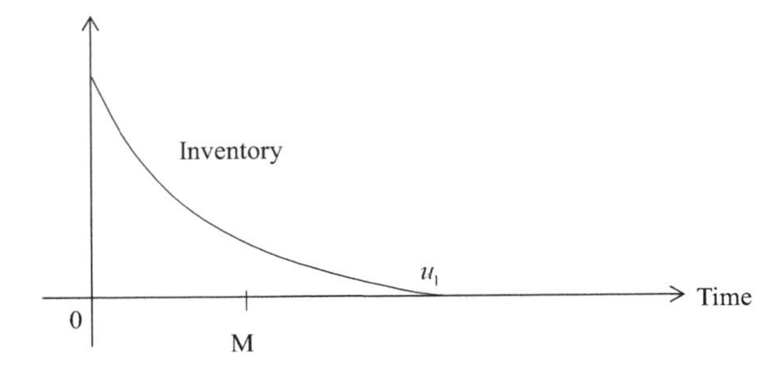


**Fig. 2.3** Inventory time graph when $M < u_1$

And interest charged is given as follows:

$$I.C_1 = 0$$

**Case 2**: When $M < u_1$ (Fig. 2.3)

For this case, vendor has to settle all his payments before to sold out all the stock. For interest earned and interest charged two following cases take the place:

**Case 2.1**: When M $< u_1$ and

$$pD\,[0, M] + I.V_{2.1}\,[0, M] \geq cE\,(0): \tag{2.23}$$

For this case, vendor has enough amount to settle all his payments. Interest charged would be zero for this case, but interest would be earned in the duration $[0, M]$.

$$I.C_{2.1} = 0 \tag{2.24}$$

$$I.V_{2.1} = pI_e \int_0^M (\delta + \beta t + \gamma E(t))t\mathrm{d}t \tag{2.25}$$

$$\begin{aligned} = pI_e\Bigg\{ & \frac{\delta M^2}{2} + \frac{\beta M^3}{3} - \gamma\Bigg(\frac{\delta(k+\gamma)+\beta}{8}M^4 \\ & + \frac{\delta M^3}{6} + \frac{\beta(k+\gamma)}{10}M^5 - \frac{\delta(k+\gamma)}{12}M^4 \\ & - \frac{(\delta(k+\gamma)^2 + \beta(k+\gamma))}{15}M^5 - \frac{\beta(k+\gamma)^2}{9}M^6\Bigg)\Bigg\} \end{aligned} \tag{2.26}$$

**Case 2.2**: When $M < u_1$ and

$$pD\,[0, M] + I.V_{2.2}\,[0, M] < cE\,(0): \tag{2.27}$$

For this case, vendor has not enough amount to settle all his payments so interest would be charged on unpaid amount. In the duration $[0, M]$ earned interest is given by as follows:

$$I.V_{2.2} = pI_e \int_0^M (\delta + \beta t + \gamma E(t))t\,\mathrm{d}t$$
$$= pI_e \left\{ \frac{\delta M^2}{2} + \frac{\beta M^3}{3} - \gamma \left( \frac{\delta(k+\gamma)+\beta}{8} M^4 + \frac{\delta M^3}{6} + \frac{\beta(k+\gamma)}{10} M^5 - \frac{\delta(k+\gamma)}{12} M^4 - \frac{(\delta(k+\gamma)^2 + \beta(k+\gamma))}{15} M^5 - \frac{\beta(k+\gamma)^2}{9} M^6 \right) \right\} \tag{2.28}$$

Interest charged on unpaid amount is given by as follows:

$$I.C_{2.2} = B.I_c \tag{2.29}$$

$$B = cE_1(0) - \{pD\,[0, M] + I.V_{2.2}\,[0, M]\} \tag{2.30}$$

$$= \left\{ \left[ c\left( \delta u_1 + \frac{\beta u_1^2}{2} + (k+\gamma)\left( \frac{\delta u_1^2}{2} + \frac{\beta u_1^3}{3} \right) \right) \right] - p\left[ \delta M + \frac{\beta M^2}{2} - \gamma \left( \frac{\delta M^2}{2} + \frac{\delta(k+\gamma)+\beta}{3} M^3 + \frac{\beta(k+c)}{4} M^4 - \delta(k+\gamma)\frac{M^3}{6} - \frac{(\delta(k+\gamma)^2 + \beta(k+\gamma))}{8} M^4 \right) - \frac{\beta(k+\gamma)^2}{10} M^5 \right) \right]$$
$$- pI_e \left[ \frac{\delta M^2}{2} + \frac{\beta M^3}{3} - \gamma \left( \frac{\delta(k+\gamma)+\beta}{8} M^4 + \frac{\delta M^3}{6} + \frac{\beta(k+\gamma)}{10} M^5 - \frac{\delta(k+\gamma)}{12} M^4 - \frac{(\delta(k+\gamma)^2 + \beta(k+\gamma))}{15} M^5 - \frac{\beta(k+\gamma)^2}{9} M^6 \right) \right] \right\}$$

**Case 3**: When cash discount facility is given:

For this case, retailer provides cash discount at a rate of $y\%$ to settle all his dues at the arrival of the stock. Interest earn would be

$$I.V_3 = pI_e \int_0^T (\delta + \beta t + \gamma E(t))\mathrm{d}t \tag{2.31}$$

$$pI_e\left\{\delta T + \frac{\beta T^2}{2} - c\left(\frac{\delta u_1^2}{2} + \frac{\delta(k+\gamma)+\beta}{3}u_1^3 + \frac{\beta(k+\gamma)}{4}u_1^4 - \delta(k+\gamma)\frac{u_1^3}{6} - \frac{(\delta(k+\gamma)^2+\beta(k+\gamma))}{8}u_1^4\right)\right\}$$

Purchasing cost for this case would be

$$P.C_x = \left\{\delta u_1 + \beta\frac{u_1^2}{2} + (k+\gamma)\left(\delta\frac{u_1^2}{2} + \beta\frac{u_1^2}{3}\right) + \frac{\delta}{2}(T^2 - u_1^2) + \frac{\beta}{3}(T^3 - u_1^3)\right\}c\left(1 - \frac{y}{100}\right) \tag{2.32}$$

## 2.7 Unit Time Profit

Unit time profit for the system is given by as follows:

$$U.T.P_x = \frac{1}{T}\{S.R_x. - P.C_x. - H.C_x. - D.C_x. - L.S.C_x. - S.C_x. - O.C_x. - I.C. + I.V\} \tag{2.33}$$

## 2.8 Numerical Example

**Case 1**: When $M \geq u_1$

$A = 300$ per/order, $c = 22$ Rs/unit, $d = 21$, $k = 0.001$, $T = 28$ days, $M = 22$ days, $\delta = 300$ units, $\beta = 0.1$, $\gamma = 0.01$, $l_r = 7$ Rs/unit, $s_r = 5$ Rs/unit, $h_f = 0.22$ Rs/unit, $h_g = 0.15$ Rs/unit, $p = 30$ Rs/unit, $I_e = 0.02$, $n = 2$, $\lambda = 0.1$.

After solving Eq. (2.33) with the help of corresponding parameters optimal value of $u_1 = 19.6461$ days and $U.T.P_x = 2077.92$ Rs. and optimal ordered quantity $Q = 8702.17$ units.

The behavior of the system for $U.T.P_x$. is given by Figs. 2.4 and 2.5 with the help of Mathematica 11.3.

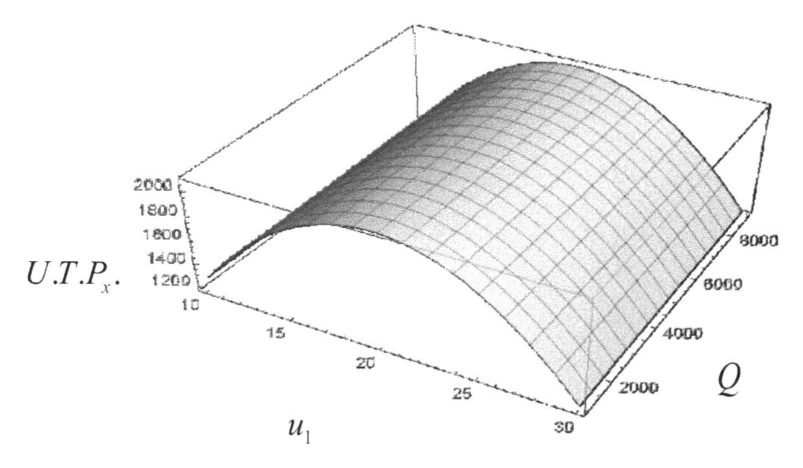


**Fig. 2.4** Behavior of $U.T.P_x$. with respect to $u_1$ and $Q$

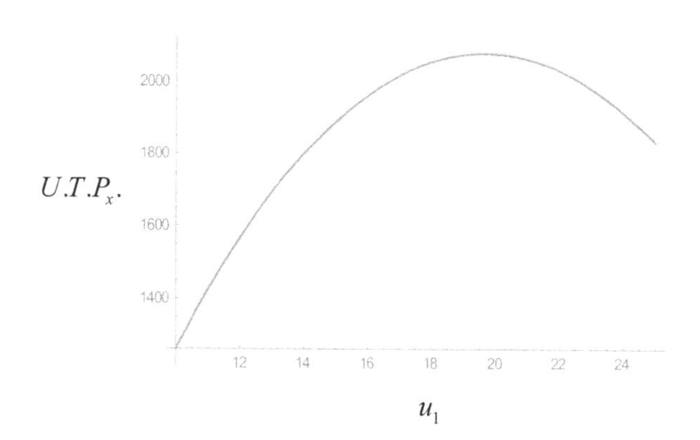


**Fig. 2.5** Behavior of $U.T.P_x$. with rest to $u_1$

**Case 2.1**: When $M < u_1$ and

$$pD\,[0, M] + IV_{2.1}\,[0, M] \geq cE\,(0):$$

$A = 300$ per/order, $c = 22$ Rs/unit, $d = 21$, $k = 0.001$, $T = 28$ days, $M = 17$ days, $\delta = 300$ units, $\beta = 0.1$, $\gamma = 0.01$, $l_r = 7$ Rs/unit, $s_r = 5$ Rs/unit, $h_f = 0.22$ Rs/unit, $h_g = 0.15$ Rs/unit, $p = 30$ Rs/unit, $I_e = 0.02$, $n = 2$, $\lambda = 0.1$.

After solving Eq. (2.33) with the help of corresponding parameters optimal value of $u_1 = 18.5963$ days and $U.T.P_x = 1533.71$ Rs. and optimal ordered quantity $Q = 8535.0$ units.

The behavior of the system for $U.T.P_x$. is given by Figs. 2.6 and 2.7 with the help of Mathematica 11.3.

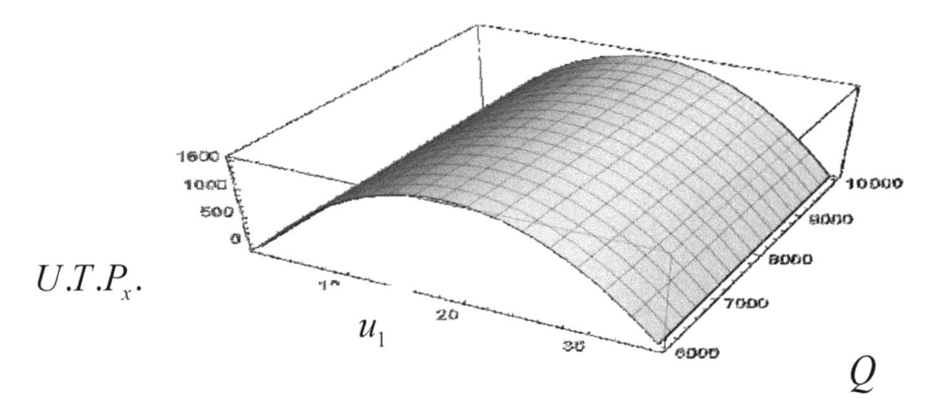


**Fig. 2.6** Behavior of $U.T.P_x.$ with respect to $u_1$ and $Q$

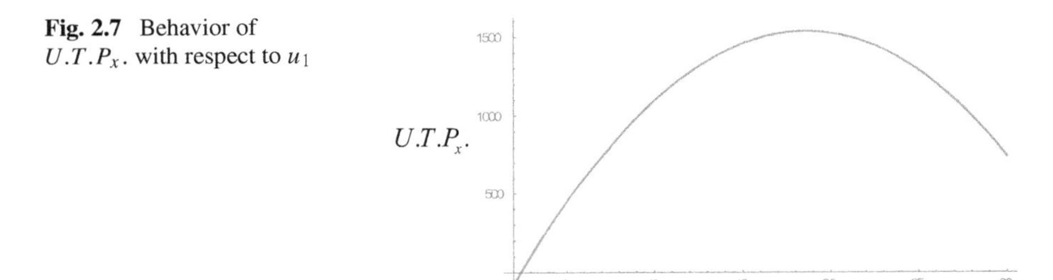


**Fig. 2.7** Behavior of $U.T.P_x.$ with respect to $u_1$

**Case 2.2**: When M $< u_1$ and

$$pD\,[0, M] + IV_{2.2}\,[0, M] < cE\,(0):$$

$A = 300$ per/order, $c = 22$ Rs/unit, $d = 21$, $k = 0.001$, $T = 28$ days, $M = 17$ days, $\delta = 300$ units, $\beta = 0.1$, $\gamma = 0.01$, $l_r = 7$ Rs/unit, $s_r = 5$ Rs/unit, $h_f = 0.22$ Rs/unit, $h_g = 0.15$ Rs/unit, $p = 30$ Rs/unit, $I_e = 0.02$, $n = 2$, $\lambda = 0.1$, $I_c = 0.016$.

After solving Eq. (2.33) with the help of corresponding parameters optimal value of $u_1 = 18.5961$ days and $U.T.P_x = 1627.54$ Rs. and optimal ordered quantity $Q = 8534.97$ units.

The behavior of the system for $U.T.P_x.$ is given by Figs. 2.8 and 2.9 with the help of Mathematica 11.3.

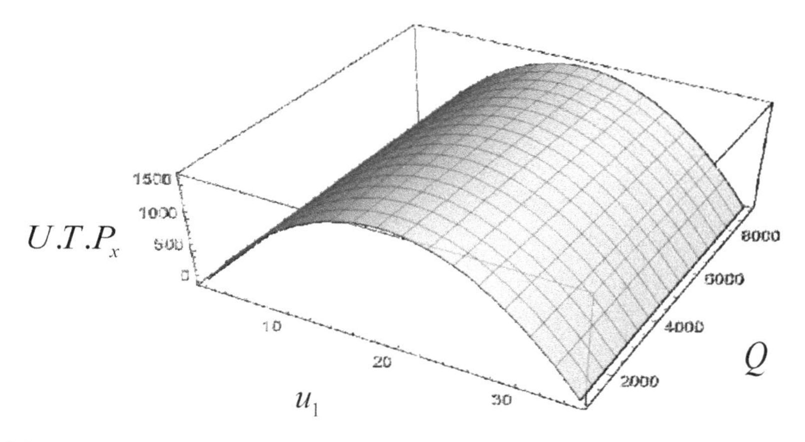


**Fig. 2.8** Behavior of $U.T.P_x$. with respect to $u_1$ and $Q$

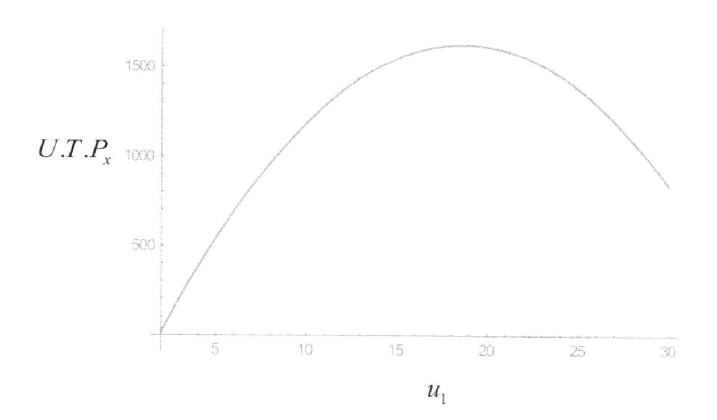


**Fig. 2.9** Behavior of $U.T.P_x$. with respect to $u_1$

**Case 3**: When cash discount facility is given:

$A = 300$ per/order, $c = 22$ Rs/unit, $d = 21$, $k = 0.001$, $T = 28$ days, $\delta = 300$ units, $\beta = 0.1$, $\gamma = 0.01$, $l_r = 7$ Rs/unit, $s_r = 5$ Rs/unit, $h_f = 0.22$ Rs/unit, $h_g = 0.15$ Rs/unit, $p = 30$ Rs/unit, $I_e = 0.02$, $n = 2$, $\lambda = 0.1$, $y = 0.02$.

After solving this model with the help of corresponding parameters optimal value of $u_1 = 18.4906$ days and $U.T.P_x = 826.307$ Rs. and optimal ordered quantity $Q = 8517.72$ units.

The behavior of the system for $U.T.P_x$. is given by Figs. 2.10 and 2.11 with the help of Mathematica 11.3.

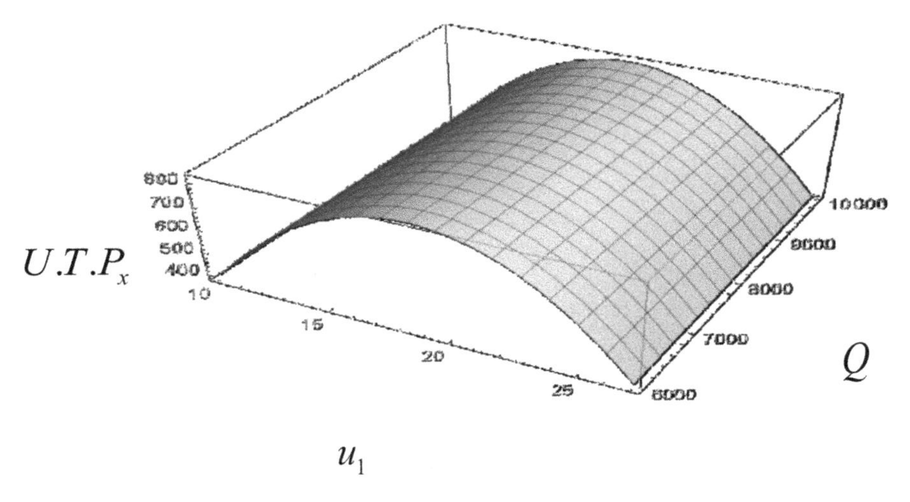


**Fig. 2.10** Behavior of $U.T.P_x$. with respect to $u_1$ and $Q$

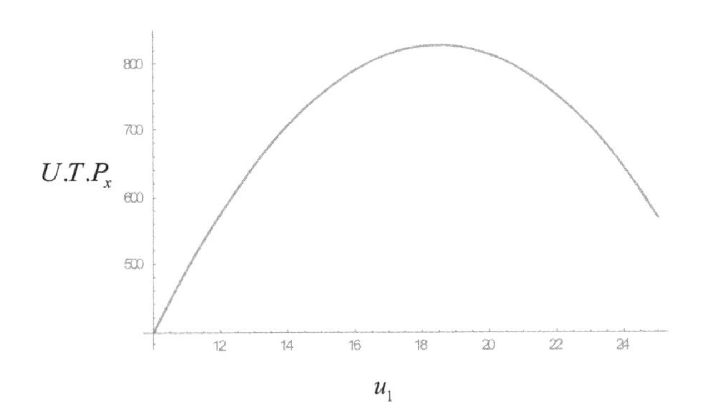


**Fig. 2.11** Behavior of $U.T.P_x$. with respect to $u_1$

## 2.9 Sensitivity Analysis

Sensitivity analysis for distinct parameters are specified as follows:

**Case 1**: When M $\geq u_1$ (Tables 2.1, 2.2, 2.3, 2.4, 2.5 and 2.6).

**Case 2**: When $M < u_1$ (Tables 2.7, 2.8, 2.9, 2.10, 2.11 and 2.12).

## 2.10 Observations

- Tables 2.1 and 2.7 represent the effect of a on $u_1$ and on $U.T.P_r$, it is observed that after an increment in a, value of $u_1$ in both the tables remain unaffected while

**Table 2.1** Variation in optimal solution for demand parameter ($\delta$)

| % change in ($\delta$) (%) | ($\delta$) | $u_1$ | $U.T.P_r.$ |
|---|---|---|---|
| −20 | 240 | 19.6461 | 2080.06 |
| −15 | 255 | 19.6461 | 2079.52 |
| −10 | 270 | 19.6461 | 2078.99 |
| −5 | 285 | 19.6461 | 2078.45 |
| 0 | 300 | 19.6461 | 2077.92 |
| 5 | 315 | 19.6461 | 2077.38 |
| 10 | 330 | 19.6461 | 2076.85 |
| 15 | 345 | 19.6461 | 2076.31 |
| 20 | 360 | 19.6461 | 2075.77 |

**Table 2.2** Variation in optimal solution for shortage parameter ($s_r$)

| % change in ($s_r$) (%) | ($s_r$) | $u_1$ | $U.T.P_r.$ |
|---|---|---|---|
| −20 | 4 | 19.021 | 2171.51 |
| −15 | 4.25 | 19.1771 | 2147.48 |
| −10 | 4.5 | 19.3332 | 2123.87 |
| −5 | 4.75 | 19.4896 | 2100.68 |
| 0 | 5 | 19.6461 | 2077.92 |
| 5 | 5.25 | 19.8027 | 2055.57 |
| 10 | 5.5 | 19.9595 | 2033.65 |
| 15 | 5.75 | 20.1165 | 2012.16 |
| 20 | 6 | 20.2737 | 1991.08 |

**Table 2.3** Variation in optimal solution for lost sale cost parameter ($l_r$)

| % change in ($l_r$) (%) | ($l_r$) | $u_1$ | $U.T.P_r.$ |
|---|---|---|---|
| −20 | 5.6 | 19.3762 | 2097.36 |
| −15 | 5.95 | 19.4453 | 2092.38 |
| −10 | 6.3 | 19.5132 | 2087.48 |
| −5 | 6.65 | 19.5802 | 2082.66 |
| 0 | 7 | 19.6461 | 2077.92 |
| 5 | 7.35 | 19.711 | 2073.25 |
| 10 | 7.7 | 19.7749 | 2068.65 |
| 15 | 8.05 | 19.8379 | 2064.12 |
| 20 | 8.4 | 19.8999 | 2059.66 |

some decrement in $U.T.P_r$ in Table 2.1 and some increment in $U.T.P_r$ in Table 2.7 are detected.

**Table 2.4** Variation in optimal solution for deterioration cost parameter $(d)$

| % change in $(d)$ (%) | $(d)$ | $u_1$ | $U.T.P_r.$ |
|---|---|---|---|
| −20 | 16.8 | 20.2326 | 2176.73 |
| −15 | 17.85 | 20.0825 | 2151.48 |
| −10 | 18.9 | 19.9347 | 2126.6 |
| −5 | 19.95 | 19.7893 | 2102.08 |
| 0 | 21 | 19.6461 | 2077.92 |
| 5 | 22.05 | 19.505 | 2054.1 |
| 10 | 23.1 | 19.378 | 2032.63 |
| 15 | 24.15 | 19.2293 | 2007.49 |
| 20 | 25.2 | 19.0946 | 1984.67 |

**Table 2.5** Variation in optimal solution for deterioration parameter $(k)$

| % change in $(k)$ (%) | $(k)$ | $u_1$ | $U.T.P_r.$ |
|---|---|---|---|
| −20 | 0.00096 | 18.3506 | 1869.34 |
| −15 | 0.00102 | 18.2662 | 1855.59 |
| −10 | 0.00108 | 18.1827 | 1841.98 |
| −5 | 0.00114 | 18.1002 | 1828.50 |
| 0 | 0.0012 | 19.6461 | 2077.92 |
| 5 | 0.00126 | 19.636 | 2076.31 |
| 10 | 0.00132 | 19.6259 | 2074.71 |
| 15 | 0.00138 | 19.6158 | 2073.1 |
| 20 | 0.00144 | 19.6058 | 2071.5 |

**Table 2.6** Variation in optimal solution for interest earned parameter $(I_e)$

| % change in $(I_e)$ (%) | $(I_e)$ | $u_1$ | $U.T.P_r.$ |
|---|---|---|---|
| −20 | 0.016 | 19.4902 | 1729.38 |
| −15 | 0.017 | 19.531 | 1863.73 |
| −10 | 0.018 | 19.5705 | 1935.1 |
| −5 | 0.019 | 19.6089 | 2006.5 |
| 0 | 0.02 | 19.6461 | 2077.92 |
| 5 | 0.021 | 19.6822 | 2149.36 |
| 10 | 0.022 | 19.7172 | 2220.83 |
| 15 | 0.023 | 19.7513 | 2292.32 |
| 20 | 0.024 | 19.7844 | 2363.83 |

**Table 2.7** Variation in optimal solution for demand parameter ($\delta$)

| % change in ($\delta$) (%) | ($\delta$) | $u_1$ | $U.T.P_r.$ |
|---|---|---|---|
| −20 | 240 | 18.5961 | 1300.82 |
| −15 | 255 | 18.5961 | 1382.54 |
| −10 | 270 | 18.5961 | 1464.21 |
| −5 | 285 | 18.5961 | 1545.87 |
| 0 | 300 | 18.5961 | 1627.54 |
| 5 | 315 | 18.5961 | 1709.2 |
| 10 | 330 | 18.5961 | 1790.87 |
| 15 | 345 | 18.5961 | 1872.53 |
| 20 | 360 | 18.5961 | 1954.2 |

**Table 2.8** Variation in optimal solution for lost sale cost parameter ($l_r$)

| % change in ($l_r$) (%) | ($l_r$) | $u_1$ | $U.T.P_r.$ |
|---|---|---|---|
| −20 | 5.6 | 18.1531 | 1652.52 |
| −15 | 5.95 | 18.2678 | 1646.06 |
| −10 | 6.3 | 18.3798 | 1639.74 |
| −5 | 6.65 | 18.4892 | 1633.57 |
| 0 | 7 | 18.5961 | 1627.54 |
| 5 | 7.35 | 18.7006 | 1621.64 |
| 10 | 7.7 | 18.8027 | 1615.87 |
| 15 | 8.05 | 18.9111 | 1609.75 |
| 20 | 8.4 | 19.0003 | 1604.7 |

**Table 2.9** Variation in optimal solution for deterioration cost parameter ($d$)

| % change in ($d$) (%) | ($d$) | $u_1$ | $U.T.P_r.$ |
|---|---|---|---|
| −20 | 16.8 | 19.4008 | 1717.21 |
| −15 | 17.85 | 19.1934 | 1694.07 |
| −10 | 18.9 | 18.9902 | 1671.42 |
| −5 | 19.95 | 18.7911 | 1649.25 |
| 0 | 21 | 18.5961 | 1627.54 |
| 5 | 22.05 | 18.4049 | 1606.27 |
| 10 | 23.1 | 18.2175 | 1585.44 |
| 15 | 24.15 | 18.0337 | 1565.03 |
| 20 | 25.2 | 17.8535 | 1545.03 |

**Table 2.10** Variation in optimal solution for shortage cost parameter $(s_r)$

| % change in $(s_r)$ (%) | $(s_r)$ | $u_1$ | $U.T.P_r.$ |
|---|---|---|---|
| −20 | 4 | 17.6956 | 1733.93 |
| −15 | 4.25 | 17.921 | 1706.42 |
| −10 | 4.5 | 18.1462 | 1679.52 |
| −5 | 4.75 | 18.3712 | 1653.23 |
| 0 | 5 | 18.5961 | 1627.54 |
| 5 | 5.25 | 18.8208 | 1602.46 |
| 10 | 5.5 | 19.0452 | 1577.11 |
| 15 | 5.75 | 19.2696 | 1554.11 |
| 20 | 6 | 19.4937 | 1530.84 |

**Table 2.11** Variation in optimal solution for deterioration parameter $(k)$

| % change in $(k)$ (%) | $(k)$ | $u_1$ | $U.T.P_r.$ |
|---|---|---|---|
| −20 | 0.00096 | 19.1732 | 1692.05 |
| −15 | 0.00102 | 18.9201 | 1675.32 |
| −10 | 0.00108 | 18.5521 | 1674.23 |
| −5 | 0.00114 | 18.4421 | 1652.25 |
| 0 | 0.0012 | 18.5961 | 1627.54 |
| 5 | 0.00126 | 18.4121 | 1620.02 |
| 10 | 0.00132 | 18.2121 | 1518.20 |
| 15 | 0.00138 | 18.0121 | 1515.12 |
| 20 | 0.00144 | 17.4224 | 1512.12 |

**Table 2.12** Variation in optimal solution for interest earned parameter $(I_e)$

| % change in $(I_e)$ (%) | $(I_e)$ | $u_1$ | $U.T.P_r$ |
|---|---|---|---|
| −20 | 0.016 | 18.5961 | 1449.16 |
| −15 | 0.017 | 18.5961 | 1493.75 |
| −10 | 0.018 | 18.5961 | 1538.35 |
| −5 | 0.019 | 18.5961 | 1582.94 |
| 0 | 0.02 | 18.5961 | 1627.54 |
| 5 | 0.021 | 18.5961 | 1672.13 |
| 10 | 0.022 | 18.5961 | 1716.73 |
| 15 | 0.023 | 18.5961 | 1761.32 |
| 20 | 0.024 | 18.5961 | 1805.92 |

- Tables 2.2 and 2.10 represent the effect of $s_r$ on $u_1$ and on $U.T.P_r$, it is observed that after an increment in $s_r$, some increment in $u_1$ and some decrement in $U.T.P_r$ in both the tables are detected.
- Tables 2.3 and 2.8 represent the effect of $l_r$ on $u_1$ and on $U.T.P_r$, it is observed that after an increment in $l_r$, some increment in $u_1$ and some decrement in $U.T.P_r$ in both the tables are detected.
- Tables 2.4 and 2.9 represent the effect of d on $u_1$ and on $U.T.P_r$, it is observed that after an increment in d, some decrement in $u_1$ and $U.T.P_r$ in both the tables are detected.
- Tables 2.5 and 2.11 represent the effect of k on $u_1$ and on $U.T.P_r$, it is observed that after an increment in k, some increment in $u_1$ and $U.T.P_r$ in Table 2.5 while some decrement in $u_1$ and $U.T.P_r$ in Table 2.11 are detected.
- Tables 2.6 and 2.12 represent the effect of $I_e$ on $u_1$ and on $U.T.P_r$, it is observed that after an increment in $I_e$, value of $u_1$ remains unaffected in Table 2.12 while some increment in $u_1$ in Table 2.6 and $U.T.P_r$ in both the tables are detected.

## 2.11 Conclusions

Present paper is concerned with inventory policies for variable demand under some real-life situations like cash discount and learning effect. Shortages are also allowed with partial backlogging and backlogging rate present in the model is assumed as a waiting time-dependent function. All these facts together make this study very unique and straight forward. To improve the efficiency of the model numerical examples for different cases and sensitivity analysis for distinct value of parameters have been discussed with the help of Mathematica 11.3. This Model further can be modified for different demands, deterioration, and more cases of backlogging rate. Also, can be extended for different realistic approaches such as inflationary environment and preservation technology.

## References

Bardhan S, Pal H, Giri BC (2019) Optimal replenishment policy and preservation technology investment for a non-instantaneous deteriorating item with stock-dependent demand. Oper Res Int J 19(2):347–368

Chowdhury RR, Ghosh SK, Chaudhuri KS (2014) An inventory model for deteriorating items with stock and price sensitive demand. Int J Appl Comput Math 1(2):187–201

Geetha KV, Udayakumar R (2016) Optimal lot sizing policy for non-instantaneous deteriorating items with price and advertisement dependent demand under partial backlogging. Int J Appl Comput Math 2(2):171–193

Giri BC, Pal H, Maiti T (2017) A vendor buyer supply chain model for time dependent deteriorating items with preservation technology investment. Oper Res Int J 10(4):431–449

Goyal SK, Singh SR, Yadav D (2017) Economic quantity model for imperfect lot with partial backlogging under the effect of learning and advertisement dependent imprecise demand. Int J Oper Res 29(2):197–218
Handa N, Singh SR, Punetha N, Tayal S (2020) A trade credit policy in an EOQ model with stock sensitive demand and shortages for deteriorating items. Int J Serv Oper Inf 10(4):350–365
Khanna A, Gautam P, Jaggi CK (2017) Inventory Modelling for deteriorating Imperfect quality items with selling price dependent demand and shortage backordering under credit financing. Int J Math Eng Manag Sci 2(2):110–124
Khurana D, Chaudhary R (2016) Optimal pricing and ordering policy for deteriorating items with stock product and price dependent demand and partial backlogging. Int J Math Oper Res 12(3):331–349
Khurana D, Chaudhary RR (2018) An order level inventory model for deteriorating stock product and time dependent demand under shortages. Int J Math Oper Res 12(3):331–349
Kumar N, Singh SR (2014) Effect of salvage value on a two-warehouse inventory model for deteriorating items with stock-dependent demand and partial backlogging. Int J Oper Res 19(4):479–496
Kumar S, Kumar A, Jain M (2020) Learning effect on an optimal policy for mathematical inventory model for decaying items under preservation technology with the environment of Covid-19 pandemic. Malaya J Matematik 8(4):1694–1702
Mahapatra GS, Adak S, Mandal TK, Pal S (2017) Inventory model for deteriorating items with reliability dependent demand and partial backorder. Int J Oper Res 29(3):344–359
Rastogi M, Khushwah P, Singh SR (2018) An inventory model for non-instantaneous deteriorating products having price sensitive demand and partial backlogging of occurring shortages. Int J Oper Quant Manag 24(1):59–73
Roy T, Chaudhuri KS (2011) An inventory model for Weibull distribution deterioration under-price dependent demand and partial backlogging with opportunity cost due to lost sales. Int J Model Ident Control 13(1):56–66
Shaikh AA (2017) An inventory model for deteriorating item with frequency of advertisement and selling price dependent demand under mixed type trade credit policy. Int J Logistic Syst Manag 28(3):375–395
Shaikh AA, Cárdenas-Barrón LE, Bhunia AK, Tiwari S (2019) Inventory model for a three Weibull distributed items with variable demand dependent on price and frequency of advertisement under trade credit. RAIRO—Oper Res 53(3):903–916
Singh S, Khurana D, Tayal S (2016) An Economic order quantity model for deteriorating products having stock dependent demand with trade credit period and preservation technology. Uncertain Supply Chain Manag 4(1):29–42
Singh SR, Rathore H (2016) Reverse logistic model for deteriorating items with preservation technology investment and learning effect in an inflationary environment. Control Cybern 45(1):83–94
Singh SR, Jain S, Pareek S (2013) an imperfect quality items with learning and inflation under limited storage capacity. Int J Ind Eng Comput 4(4):479–490
Singh SR, Sharma S, Kumar M (2020) A reverse logistic model for decaying items with variable production and remanufacturing incorporating learning effect. Int J Oper Res 38(2):422–448
Skouri K, Konstantaras I, Papachristos S, Ganas I (2009) Inventory models with ramp type demand rate, partial backlogging and Weibull deterioration rate. Euro J Oper Res 192(1):79–92
Tripathi RP, Singh D, Aneja S (2018) Inventory model for stock dependent demand and time varying holding cost under different trade credits. Yugoslav J Oper Res 28(1):139–151

# Chapter 3
# Impact of Inflation on Production Inventory Model with Variable Demand and Shortages

**Nidhi Handa, S. R. Singh, and Neha Punetha**

**Abstract** The collaboration of inflation or reverse money on the production inventory system is one of the major key factors for successful supply chain management. Here, demand rate of items is increasing with time and decreasing with proportion of selling price which is effective strategy for the market change. The developed model designed for shortages is considered partially backlogged where backlogging rate is decreasing with waiting time. Considering inflation on various costs is providing more reliable result due to real-life problem. This study presents the impact of inflation on production inventory model for deteriorating items with time and selling price-dependent demand under shortages. After that, to illustrate the model, numerical example is provided and solved. At the end, sensitivity analysis is introduced to show the validity and optimality of proposed study in order to analyse the effect of changes of different key parameters.

**Keywords** Production · Deterioration · Variable demand · Shortage · Partial backlogging · Inflation

## 3.1 Introduction

In production inventory system, the aim of manufacturer is to calculate the optimal production cost with lowest price of goods. There are many factors such as demand, production, deterioration and backlogging included in the development of inventory models. Hsieh and Dye (2013) addressed the impact of preservation technology on production model having fluctuating demand with time. Majumder et al. (2015) presented a partial delay payment policy on EPQ model under crisp and fuzzy domain with declining market demand. A manufacturing supply chain model with two-level trade credit policies was proposed by Kumar et al. (2015) under the effect of learning

N. Handa · N. Punetha (✉)
Department of Mathematics and Statistics, KGC, Gurukul Kangri Vishwavidyalaya, Haridwar, India

S. R. Singh
Department of Mathematics, CCS University, Meerut, India



N. H. Shah et al. (eds.), *Decision Making in Inventory Management*,
Inventory Optimization, https://doi.org/10.1007/978-981-16-1729-4_3

and preservation technology. Mishra (2016) investigated order-level inventory model with quadratic demand rate and Weibull deterioration rate under partial backlogging. Panda et al. (2018) studied two-warehouse optimal model of decaying products having variable demand and shortages with permissible delay approach.

In past study, researcher assumed that the demand to be constant or fixed may mislead the result. Now researcher is focusing on variable demand as like stock dependent, price dependent, time dependent, etc. which provides accurate optimum solution. Sometimes, costumer compromises with the quality of items they purchase because of high price. Thus, it is very challenging task for production manufacturer to provide good-quality products in a suitable selling price. Generally, high selling price decreases customer's demand. A production model for infinite time horizon with time-dependent deterioration and price-sensitive demand has been investigated by Sharma et al. (2015). An inventory model including non-instantaneous deterioration with learning effect was studied by Shah and Naik (2018) in which assumed demand is price dependent. Also, Singh (2019) developed EPQ model with variable demand and backlogging where backlogging rate depends on waiting time.

For researcher, it is significant to include deterioration into account. In inventory models, deteriorating products are those items which we cannot use for future purpose due to decay, damage or spoilage of items. Wee and Wang (1999) introduced production policy with time-dependent demand for decaying goods. Molamohamandi et al. (2014) investigated optimal replenishment policy of EPQ model with shortages under permissible delay on payment for decaying items. Priyan and Uthayakumar (2015) studied economic manufacturing problem for defective items under imperfect production processes and reworking system. Tiwari et al. (2018) examined vendor and buyer inventory policy of imperfect products taking carbon emission into account. A lifetime decreasing model was studied by Singh and Rana (2020) with variable carrying cost and lost sale.

In the development of production models, inflation plays very important role as it affects the economy. Due to inflation, there is sustainable increase in general price or continuous fall in time value of money. Business organization is affected due to rapid inflation so it cannot be ignored. In production plants, inventories of raw materials and goods are big outlay and should be completed financially. Sarkar and Moon (2011) established manufacturing production model for imperfect items taking inflation into account. Kumar et al. (2013) studied effect of inflation on order quantity model for perishable products with lost sale. The effect of inflationary environment on ordering model under trade credit policy for finite time horizon was examined by Muniappan et al. (2015) in which deterioration is time dependent. Singh and Sharma (2016) extended production quantity model for stochastic demand and over finite time horizon. Pervin et al. (2020) developed EPQ model with collaboration of preservation technology investment to reduce deterioration rate.

In traditional models, it is assumed that shortages are completely backlogged, but it is not necessarily true in real-life scenario. Considering daily-life problems, some costumers like to wait especially in shortage period until replenishment, whereas some become restless and head elsewhere. Chang (2004) classified EPQ model for shortages using variable lead time. He also used algebraic method to find the minimum total average cost. Jain et al. (2007) extended EPQ model of decaying product for price- and stock-sensitive consumption under shortages. A non-instantaneous inventory model was proposed by Sharma and Bansal (2017) with learning effect and backlogging. An economic production quantity model for deteriorating items was studied by Khurana et al. (2018) with linear demand and partial backlogging. Handa et al. (2020) derived EOQ model under the effect of trade credit policy.

### *3.1.1 Our Contribution*

In the present study, a production inventory model for deteriorating goods with time and price sensitive demand over a finite planning horizon under the effect of inflation on costs has been established. Due to tough competition in the market, manufacturer has to provide good-quality items to the customers at lowest reasonable rate. Shortages are permitted and partially backlogged with waiting time-dependent rate. To minimize the optimal cost of the proposed model, it has been solved numerically. Furthermore, to examine the stability of this problem, sensitivity analysis has been done for different parameters. The findings suggested that the production model may reduce the total average cost by taking inflation on account.

## 3.2 Assumptions

The model is supported by the following assumptions:

1. The production cycle is for finite planning horizon.
2. The demand of the items depends on time as well as on price such as

$$D(t, s_1) = \alpha + \beta t - \gamma s_1 \quad \text{where } \alpha > \beta \text{ and } \gamma \ll 1$$

3. The model has allowed constant rate of deterioration $\sigma$.
4. Deteriorating items are not repaired during whole cycle.
5. Inflation is allowed, and lead time is zero.
6. Shortages occur and not completely backlogged.
7. Backlogging rate is assumed to be decreasing with waiting time and given by

$$\lambda(\delta) = 1 - \frac{\delta}{T}$$

8. The production cycle is for finite planning horizon.

## 3.3 Notations

| | |
|---|---|
| $Q_1(t)$ | Positive inventory level. |
| $Q_2(t)$ | Maximum inventory level when shortage occurs. |
| $K$ | Production rate for manufacturing. |
| $P_1$ | Production cost per unit. |
| $s_1$ | Selling price. |
| $A$ | Acquisition cost per unit. |
| $M$ | Setup cost per production. |
| $\sigma$ | Rate of deterioration. |
| $C_d$ | Deterioration cost per unit. |
| $C_I$ | Inventory holding cost per unit. |
| $S$ | Inventory shortage cost per unit. |
| $l$ | Cost of lost sale per unit. |
| $\alpha, \beta$ | Demand rate parameters where $\alpha > \beta$. |
| $\gamma$ | Constant function of selling price. |
| $\delta$ | Backlogging rate (constant). |
| $r$ | Constant rate of inflation. |

## 3.4 Mathematical Modelling

In Fig. 3.1, production starts from $t = 0$. At $t = 0$, there is no production and demand. Production and consumption jointly continued during $[0, u_1]$, production reaches its maximum size at time $t = u_1$, and it stops at this point. After that, due to combined effect of consumption of inventory and deterioration of items, inventory depletes zero during $[u_1, u_2]$. After that, shortages occur including partial backlogging between $t = u_2$ to $t = T$.

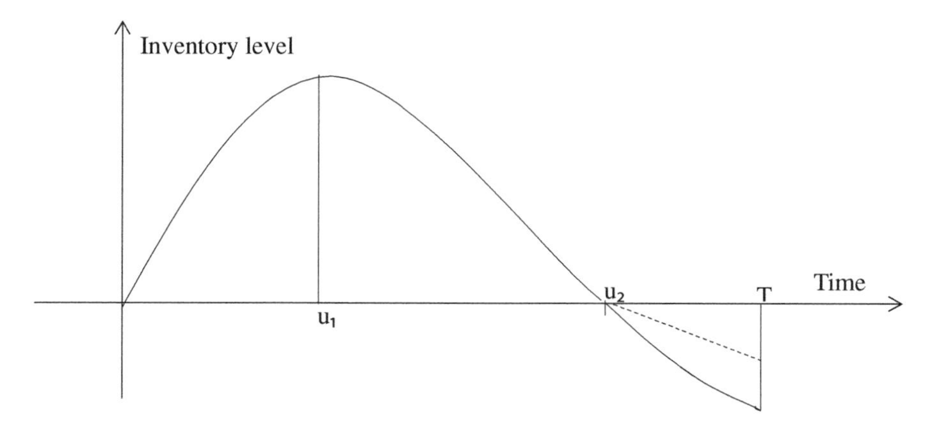


**Fig. 3.1** Inventory versus time graph

The behaviour of inventory system with time has been specified by below-given differential equations:

$$\frac{dQ_1(t)}{dt} = K - (\alpha + \beta t - \gamma s_1) - \sigma Q_1(t) \quad 0 \le t \le u_1 \tag{3.1}$$

$$\frac{dQ_2(t)}{dt} = K - (\alpha + \beta t - \gamma s_1) - \sigma Q_2(t) \quad u_1 \le t \le u_2 \tag{3.2}$$

$$\frac{dQ_3(t)}{dt} = -(\alpha + \beta t - \gamma s_1) \quad u_2 \le t \le T \tag{3.3}$$

Boundary conditions are as follows:

$$Q_1(0) = 0,\ Q_2(u_1) = 0,\ Q_3(T) = 0$$

Solution of these equations with the help of boundary conditions:

$$Q_1(t) = \left[(K - \alpha + \gamma s_1)\left(t + \frac{\sigma t^2}{2}\right) - \beta\left(\frac{t^2}{2} + \frac{\sigma t^3}{3}\right)e^{-\sigma t}\right] \tag{3.4}$$

$$Q_2(t) = \Big[(\alpha - \gamma s_1)\{(u_2 - t) + \frac{\sigma}{2}(u_2^2 + t^2)\} + \beta\left\{\left(\frac{u_2^2 - t^2}{2}\right) + \sigma\left(\frac{u_2^3 - t^3}{3}\right)\right\}\Big]e^{-\sigma t} \tag{3.5}$$

and $$Q_3(t) = \left[(\alpha - \gamma s_1)(u_2 - t) + \frac{\beta}{2}(u_2^2 - t^2)\right] \tag{3.6}$$

Maximum inventory in $[0, u_2]$

$$Q_1(u_1) = Q_m$$

$$Q_m = \left[(K - \alpha + \gamma s_1)\left(u_1 + \sigma\frac{u_1^2}{2}\right) - \beta\left(\frac{u_1^2}{2} + \frac{\sigma u_1^3}{2}\right)\right]e^{-\sigma u_1} \tag{3.7}$$

And maximum inventory in shortage period $[u_2, T]$

$$Q_3(T) = Q_s$$

$$Q_s = \left[(\alpha - \gamma s_1)(u_2 - T) + \frac{\beta}{2}(u_2^2 - T^2)\right] \tag{3.8}$$

Total ordered quantity

$$Q = Q_m + Q_s \tag{3.9}$$

By using boundary condition

$$Q_1(u_1) = Q_2(u_2)$$
$$u_1 = \frac{-1}{\sigma} + \frac{1}{\sigma}\left[1 + \frac{2\sigma}{K}(\alpha - \gamma s_1)\left(u_2 + \frac{\sigma u_2^2}{2}\right)\right]^{1/2} \tag{3.10}$$

## 3.5 Cost analysis

Production cost

$$P.C. = P_1 \int_0^{u_1} K e^{-rt}\, dt$$
$$= P_1 K\left(u_1 - \frac{r u_1^2}{2}\right) \tag{3.11}$$

Acquisition cost

$$A.C. = A \tag{3.12}$$

Setup cost

$$S.C. = M \tag{3.13}$$

Deterioration cost

$$D.C. = c_d\left[\int_0^{u_1} \sigma Q_1(t)e^{-rt}dt + \int_{u_1}^{u_2} \sigma Q_2(t)e^{-rt}dt\right]$$
$$= c_d\sigma\Bigg[(K - \alpha + \gamma s_1)\left(\frac{u_1^2}{2} + \frac{\sigma u_1^3}{6}\right) - \alpha\,(\alpha - \gamma s_1)\Bigg(\left(\frac{u_2^2}{2} + \frac{\sigma u_2^3}{3}\right)$$
$$- \left(u_2 u_1 - \frac{u_1^2}{2}\right) - \frac{\sigma}{2}\left(u_2^2 u_1 - \frac{u_1^3}{3}\right)\Bigg) + \beta\Bigg(\left(\frac{u_2^3}{3} + \frac{\sigma u_2^4}{4}\right)$$
$$- \frac{1}{2}\left(u_2^2 u_1 - \frac{u_1^3}{3}\right) + \frac{\sigma}{3}\left(u_2^3 u_1 - \frac{u_1^4}{4}\right) - \left(\frac{u_1^3}{6} + \frac{\sigma u_1^4}{12}\right)\Bigg)$$
$$- (r + \sigma)\Bigg((K - \alpha + \gamma s_1)\left(\frac{u_1^3}{3} + \frac{\sigma u_1^4}{8}\right) - \beta\left(\frac{u_1^4}{8} + \frac{\sigma u_1^5}{5}\right)$$

$$+(\alpha-\gamma s_1)\left(\left(\left(\frac{u_2^3}{6}+\frac{\sigma u_2^4}{8}\right)-\left(\frac{u_2u_1^2}{2}-\frac{u_1^3}{3}\right)-\frac{\sigma}{2}\left(\frac{u_2^2u_1^2}{2}-\frac{u_1^4}{4}\right)\right)\right.$$

$$\left.\left.\left.+\beta\left(\frac{u_2^4}{8}+\frac{u_2^5}{10}\right)-\frac{1}{2}\left(\frac{u_2^2u_1^2}{2}-\frac{u_1^4}{4}\right)-\frac{\sigma}{3}\left(\frac{u_2^3u_1^2}{2}-\frac{u_1^5}{5}\right)\right)\right)\right] \tag{3.14}$$

Holding Cost

$$H.C.=c_1\left[\int_0^{u_1}Q_1(t)\,\mathrm{e}^{-rt}\mathrm{d}t+\int_{u_1}^{u_2}Q_2(t)\,\mathrm{e}^{-rt}\mathrm{d}t\right]$$

$$=c_1\left[(K-\alpha+\gamma s_1)\left(\frac{u_1^2}{2}+\frac{\sigma u_1^3}{6}\right)-\alpha(\alpha-\gamma s_1)\left(\left(\frac{u_2^2}{2}+\frac{\sigma u_2^3}{3}\right)\right.\right.$$

$$\left.-\left(u_2u_1-\frac{u_1^2}{2}\right)-\frac{\sigma}{2}\left(u_2^2u_1-\frac{u_1^3}{3}\right)\right)+\beta\left(\left(\frac{u_2^3}{3}+\frac{\sigma u_2^4}{4}\right)-\frac{1}{2}\left(u_2^2u_1-\frac{u_1^3}{3}\right)\right.$$

$$\left.\left.+\frac{\sigma}{3}\left(u_2^3u_1-\frac{u_1^4}{4}\right)-\left(\frac{u_1^3}{6}+\frac{\sigma u_1^4}{12}\right)\right)-(r+\sigma)\left((K-\alpha+\gamma s_1)\left(\frac{u_1^3}{3}+\frac{\sigma u_1^4}{8}\right)\right.$$

$$-\beta\left(\frac{u_1^4}{8}+\frac{\sigma u_1^5}{5}\right)+(\alpha-\gamma s_1)\left(\left(\left(\frac{u_2^3}{6}+\frac{\sigma u_2^4}{8}\right)-\left(\frac{u_2u_1^2}{2}-\frac{u_1^3}{3}\right)-\frac{\sigma}{2}\left(\frac{u_2^2u_1^2}{2}-\frac{u_1^4}{4}\right)\right)\right)$$

$$\left.\left.\left.+\beta\left(\frac{u_2^4}{8}+\frac{u_2^5}{10}\right)-\frac{1}{2}\left(\frac{u_2^2u_1^2}{2}-\frac{u_1^4}{4}\right)-\frac{\sigma}{3}\left(\frac{u_2^3u_1^2}{2}-\frac{u_1^5}{5}\right)\right)\right)\right] \tag{3.15}$$

Shortage cost

$$S.C.=s\int_{u_2}^{T}(\alpha+\beta t-\gamma s_1)\mathrm{e}^{-rt}\mathrm{d}t$$

$$=s\,[\,(\alpha-\gamma s_1)(T-u_2)+\frac{\beta}{2}(T^2-u_2^2)$$

$$-\,r\,\{\,\frac{(\alpha-\gamma s_1)}{2}\,(\,T^2-u_2^2)+\,\frac{\beta}{3}(\,T^3-u_2^3)\}] \tag{3.16}$$

Lost sale cost

$$L.S.C.=l\int_{u_2}^{T}(1-\lambda(\delta))\,(\alpha+\beta t-\gamma s_1)\,\mathrm{e}^{-rt}\mathrm{d}t$$

$$=\frac{l\delta}{T}\,[\,(\alpha-\gamma s_1)(T-u_2)+\frac{\beta}{2}(T^2-u_2^2)$$

$$-\,\mathrm{r}\,\{\,\frac{(\alpha-\gamma s_1)}{2}\,(\,T^2-u_2^2)+\,\frac{\beta}{3}(\,T^3-u_2^3)\}] \tag{3.17}$$

Now, by using Eqs. (3.11)–(3.17), the total average cost (TAC) of the proposed model is

$$\text{TAC}\ (u_2) = \frac{1}{T}\ [P.C. + \ O.C. + \ A.C. + \ D.C + \ H.C. + \ S.C. + \ L.S.C.] \tag{3.18}$$

### 3.5.1 Solution Procedure

The objective of the study is to obtain the optimal value of critical time ($u_2$) that minimizes the total average cost. Individual optimization method is used to determine the minimum cost per unit time of the proposed model. The value of optimal time will be calculated by using

$$\frac{\partial \text{TAC}}{\partial u_2} = 0 \text{ provided } \frac{\partial^2 \text{TAC}}{\partial u_2^2} > 0$$

## 3.6 Numerical Example

With the help of numerical aptitudes, the optimal value of time and optimal value of total average cost are calculated. For the calculation, mathematical software Mathematica 11.3 is used, and assigned values for parameters are as follows:

$\alpha = 1500$ unit, $\beta = 2.5$ unit, $\gamma = 0.02$ unit, $s_1 = 28$ rs/unit,

$M = 1500$ rs, $P_1 = 6$ rs/unit, $K = 1000$ unit, $r = 0.03$, $A = 10$ rs,

$c_d = 12$ rs/unit, $\sigma = 0.06$, $C_1 = 0.5$ rs/unit, $1 = 7$ rs/unit,

$s = 5$ rs/unit, $\delta = 0.8$, $T = 30$ days

By considering these values, we obtained (Fig. 3.2).

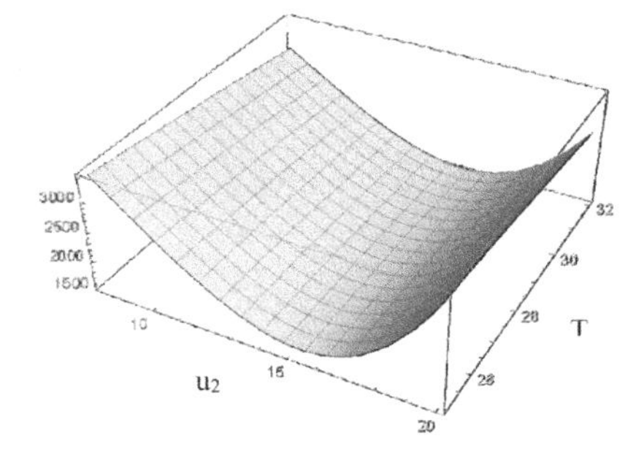


**Fig. 3.2** Convexity of total average cost with respect to $u_2$ and $T$

## 3.7 Sensitivity Analysis

Sensitive analysis is executed for the different parameters by assigning 15%, 10% and 5% decrease or an increase in each of the parameters keeping all other parameters unchanged. The variation in total average cost and critical time with the change in different parameters such as demand rate, deterioration rate, holding cost, backlogging rate and inflation rate is shown in Table 3.1.

**Table 3.1** Sensitive analysis with respect to different parameters

| Parameter | % variation | Changes in parameter | $u_2$ | TAC |
|---|---|---|---|---|
| $\beta$ | −15 | 2.125 | 15.5593 | 1497.9 |
| | −10 | 2.250 | 15.4758 | 1524.64 |
| | −5 | 2.375 | 15.3944 | 1550.76 |
| | 0 | 2.500 | 15.3151 | 1576.28 |
| | 5 | 2.625 | 15.2378 | 1601.23 |
| | 10 | 2.750 | 15.1624 | 1625.64 |
| | 15 | 2.875 | 15.0887 | 1649.53 |
| | −15 | 0.017 | 15.3159 | 1575.35 |
| | −10 | 0.018 | 15.3156 | 1575.66 |
| | −5 | 0.019 | 15.3154 | 1575.97 |
| $\gamma$ | 0 | 0.020 | 15.3151 | 1576.28 |
| | 5 | 0.021 | 15.3149 | 1576.59 |
| | 10 | 0.022 | 15.3146 | 1576.90 |
| | 15 | 0.023 | 15.3144 | 1577.21 |
| | −15 | 0.051 | 16.9367 | 1234.63 |
| | −10 | 0.054 | 16.3573 | 1357.96 |
| | −5 | 0.057 | 15.8182 | 1471.45 |
| $\sigma$ | 0 | 0.060 | 15.3151 | 1576.28 |
| | 5 | 0.063 | 14.8445 | 1673.43 |
| | 10 | 0.066 | 14.4032 | 1763.76 |
| | 15 | 0.069 | 13.9883 | 1847.99 |
| | −15 | 0.425 | 15.3477 | 1735.40 |
| | −10 | 0.450 | 15.3364 | 1682.37 |
| | −5 | 0.475 | 15.3256 | 1629.33 |
| $C_1$ | 0 | 0.500 | 15.3151 | 1576.28 |
| | 5 | 0.525 | 15.3051 | 1523.22 |
| | 10 | 0.550 | 15.2955 | 1470.15 |
| | 15 | 0.575 | 15.2862 | 1417.07 |

(continued)

**Table 3.1** (continued)

| Parameter | % variation | Changes in parameter | $u_2$ | TAC |
|---|---|---|---|---|
| | −15 | 0.680 | 15.3068 | 1569.46 |
| | −10 | 0.720 | 15.3096 | 1571.74 |
| | −5 | 0.760 | 15.3123 | 1574.01 |
| $\delta$ | 0 | 0.800 | 15.3151 | 1576.28 |
| | 5 | 0.840 | 15.3179 | 1578.55 |
| | 10 | 0.880 | 15.3207 | 1580.82 |
| | 15 | 0.920 | 15.3235 | 1583.09 |
| | −15 | 0.0255 | 15.9151 | 1885.47 |
| | −10 | 0.0270 | 15.7089 | 1786.17 |
| | −5 | 0.0285 | 15.5090 | 1683.03 |
| $r$ | 0 | 0.0300 | 15.3151 | 1576.28 |
| | 5 | 0.0315 | 15.1270 | 1466.14 |
| | 10 | 0.0330 | 14.9443 | 1352.82 |
| | 15 | 0.0345 | 14.7670 | 1236.50 |

## 3.8 Observations

1. With the increase in the value of demand parameter beta, critical time and total average cost are decreased.
2. As the coefficient of selling price gamma increases, critical time slightly decreases and total average cost slightly increases.
3. With the increment in deterioration rate, the value of critical time decreases and total average cost quickly increases.
4. On increasing the holding cost, critical time slightly decreases and total average cost decreases.
5. When the value of backlogging rate is increasing, then critical time and total average cost slightly increase.
6. When the value of inflation rate is increasing, then critical time and total average cost decrease.

## 3.9 Conclusion

In this article, the impact of inflation on production inventory model for deteriorating items has been introduced. This study contributes the idea of time- and price-sensitive demand which is a good strategy in today's competitive businesses. Due to fluctuating demand, this demand policy is suitable to fulfil customer requirements. The model follows constant deterioration which is appropriate for high deteriorating units. The organization may only save inventory production cost by proper management of deteriorating items. In practical environment, considering inflation on cost is a good policy specifically for long-term investment and forecasting. For profitable business, including shortages and backlogging provides more practical results. With the help of numerical examples, total average cost of the system has been concluded. Convexity and sensitivity have been done to check the availability and stability of the model. This approach is appropriate to meet real-life problems. For future, researcher can further extend this model for different rates of deterioration, trade credit policy, two-warehouse and variable holding cost.

## References

Chang H-C (2004) A note on the EPQ model with shortages and variable lead time. Inf Manag Sci 15(1):61–67

Handa N, Singh SR, Punetha N, Tayal S (2020) A trade credit policy in an EOQ model with stock sensitive demand and shortages for deteriorating items. Int J Serv Oper Inf 10(4):350–364

Hsieh TP, Dye CY (2013) A production- inventory model incorporating the effect of preservation technology investment when demand is fluctuating with time. J Comput Appl Math 239:25–36

Jain M, Sharma GC, Rathore S (2007) Economic production quantity model with shortages, price and stock-dependent demand for deteriorating items. Int J Eng Trans Basics 20(2):159–168

Khurana D, Tayal S, Singh SR (2018) An EPQ model for deteriorating items with variable demand rate and allowable shortages. Int J Math Oper Res 12(1):117–128

Kumar S, Handa N, Singh SR (2013) An inventory model for perishable items with partial backlogging under inflation. Int J Trends Comput Sci 2(11):551–565

Kumar S, Handa N, Singh SR, Singh D (2015) Production inventory model for low-level trade credit financing under the effect of preservation technology and learning in supply chain. Cogent Eng 2(1). https://doi.org/10.1080/23311916.2015.1045221

Majumder P, Bera UK, Maiti M (2015) An EPQ model of deteriorating items under partial trade credit financing and demand declining market in crisp and fuzzy environment. Procedia Comput Sci 45:780–789

Mishra U (2016) An EOQ model with time dependent weibull deterioration, quadratic demand and partial backlogging. Int J Appl Comput Math 2:545–563

Molamohamadi Z, Arshizadeh R, Tsmail N (2014) An EPQ inventory model with allowable shortages for deteriorating items under trade credit policy. Discrete Dyn Nat Soc 2014(6):1–10

Muniappan P, Uthayakumar R, Vanish S (2015) An EOQ model for deteriorating items with inflation and time value of money considering time- dependent deteriorating rate and delay payments. Syst Sci Control Eng 3:427–434

Panda GC, Khan MA-A, Shaikh A-A (2018) A credit policy approach in a two-warehouse inventory model for deteriorating items with price and stock- dependent demand under partial backlogging. J Ind Eng Int 15:147–170

Pervin M, Roy SK, Weber GW (2020) Deteriorating inventory with preservation technology under price and stock sensitive demand. J Ind Manage Optim 16(4):1585–1612

Priyan S, Uthayakumar R (2015) An EMQ inventory model for defective products involving rework and sales team' s initiatives- dependent demand. J Ind Eng Int 11:517–529

Sarkar B, Moon I (2011) An EPQ model with inflation in an imperfect production system. Appl Math Comput 217(13):6159–6167

Shah NH, Naik MK (2018) Inventory model for non-instantaneous deterioration and price sensitive trended demand with learning effects. Int J Inventory Res 2(1):60–77

Sharma MK, Bansal KK (2017) Inventory model for non-instantaneous decaying items with learning effect under partial backlogging and inflation. Global J Pure Appl Math 13(6):1999–2008

Sharma S, Singh SR, Ram M (2015) An EPQ model for deteriorating items with price sensitive demand and shortages. Int J Oper Res 23(2):245–255

Singh D (2019) Production inventory model of deteriorating items with holding cost, stock and selling price with backlog. Int J Math Oper Res 14(2):290–306

Singh SR, Rana K (2020) Effect of inflation and variable holding cost on life time inventory model with multi variable demand and lost sales. Int J Recent Technol Eng 8(5):5513–5519

Singh SR, Sharma S (2016) A production reliable model for deteriorating products with random demand and inflation. Int J Syst Sci 4(4):330–338

Tiwari S, Daryanto Y, Wee H-M (2018) Sustainable inventory management with deteriorating and imperfect quality items considering carbon emission. J Cleaner Production 192:281–292

Wee HM, Wang WT (1999) A variable production scheduling policy for deteriorating items with time-varying demand. Comput Oper Res 26(3):237–254

# Chapter 4
# Effect of Credit Financing on the Learning Model of Perishable Items in the Preserving Environment

**Mahesh Kumar Jayaswal, Mandeep Mittal, and Isha Sangal**

**Abstract** In this article, an EOQ mathematical model has been developed for perishable items with two level trade-credit financing policy and preservation technology under learning effect. Learning effect minimizes the holding cost and ordering cost because the holding cost and ordering cost follow the effect of the learning. Trade credit policy is very effective tool for seller to increase his sales. Furthers, seller gives a fixed credit period the buyer to increase his profit and buyer gives same policy to his customer. Some useful results determined when learning rate increases, cycle length almost fixed and retailer's total cost decreases and if trade-credit period increases then cycle length and retailer's total cost decreases due to the learning and credit financing. Finally, the total inventory cost minimizes with respect to cycle length. The numerical example describes the applicability of the present model.

**Keywords** Learning effects · EOQ · Two level trade-credit financing policy · Perishable items · Preservation · Deterioration

## 4.1 Introduction

In the field of industrial sector, there is a some problems between seller and buyer regarding optimal profit as well as total cost from the both side and such type of problems have been tried to short out by Adad and Jaggi (2003) with the help of mathematical model in which has shown the effects of credit financing policy. Shinn and Hwang (2003) calculated the maximum price for the seller as well as lot size using the policy of credit financing. The total cost acts as major role in the inventory system.

M. K. Jayaswal · I. Sangal (✉)
Department of Mathematics and Statistics, Banasthali Vidyapith, Banasthali, Rajasthan 304022, India

M. Mittal
Department of Mathematics, Amity Institute of Applied Sciences, Amity University Uttar Pradesh, Noida, Sector-125, Noida, Uttar Pradesh 201313, India



N. H. Shah et al. (eds.), *Decision Making in Inventory Management*,
Inventory Optimization, https://doi.org/10.1007/978-981-16-1729-4_4

An inventory model has been presented by Hung and Chung (2003) which is the extended form of Goyal (1985) which explained how to minimize the total cost under credit financing and payment rebate policy. A mathematical model has been developed for best costing and batch sizing in which assuming that purchase quantity and selling price both are different under credit financing policy when demand is a function of selling price. The EOQ model has been developed to easy manner as per suggested by Hung (2007) and given new idea how to find out the optimal lot size for the seller. Luo (2007) has been proposed an inventory model for the good coordination between seller and buyer under the credit financing policy. A stock model has been formulated for the optimal economic order quantity as guided by Sarmah et al. (2007) and in this model, a seller has connected to the many buyers under the credit financing to got more profit through coordinated strategy. In recent times, authors improved stock representation with the two-level credit policy. An inventory mathematical representation has been improved by Hung (2003) using the two-level credit policy and demand is stimulated as per consideration.

The research work of Huang (2003) has been investigated by Huang (2006) and improved this model by using two-level credit financing and restricted storage space. A stock model has been presented by Teng and Goyal (2007) for the customers when customer used to credit policy provided by buyer during the business. The construction of economic order quantity has been balanced by Huang (2007) which is the extended form of Haung (2003) with the help of two-level credit financing policy. An inventory model has improved for organization by Su et al. (2007) with the help of credit financing policy.

A two-level credit financing model has developed by Huang and Hsu (2008) with the help of partial credit policy. An economic order quantity formulation has been presented by Jaggi et al. (2008) under two-level credit policy with credit dependent demand. Jaber et al. (2008) has explained economic production quantity model for items with imperfect quality subjected to learning effects. Research task of the Huang (2007) have modified by Teng and Chang (2009) with the policy of two-level credit financing system which is the benefit for the customers. A model has been developed by Chen and Kang (2010) for business system under two-level credit policy with price dependent demand and conciliation situation. A lot of review article has been provided by Shah et a1. (2010) for the inventory system under two-level credit financing policy. Shah et al. (2012) explained an EOQ model in fuzzy environment and trade credit. Shah et al. (2016) presented a deteriorating inventory model under permissible delay in payments and fuzzy environment.

Wright (1936) analyzed that factor affecting the cost of airplane. The present article combined the effect of two- level trade credit and leaning effect with preservation atmosphere. Jayaswal et al. (2019) has proposed the effects of learning on retailer ordering policy for imperfect quality items with trade credit financing. Agarwal and Mittal (2019) explained an inventory classification using multilevel association rule mining. Mittal et al. (2017) represented a mathematical model on retailer's ordering policy for deteriorating imperfect quality items when demand and price are time-dependent under inflationary conditions and permissible delay in payments. Jaggi et al. (2011) proposed a credit financing for deteriorating imperfect-quality items under inflationary conditions. Jayaswal et al. (2021) proposed a mathematical model with the ordering policies for deteriorating imperfect quality items with trade-credit financing under learning effect.

## 4.2 Assumptions and Notations

The mathematical model is derived using following notations and assumptions.

### 4.2.1 Assumptions

- Replenishment rate is infinite.
- There are no shortages in this model.
- The lead-time is considered to be zero.
- Unit purchasing cost is less than the unit selling price.
- No replacement policy of perishable items during cycle length.
- Two level trade credits are allowed.

### 4.2.2 Notations

| | |
|---|---|
| $R$ | Annual constant demand. |
| $\xi$ | Preservation cost. |
| $A$ | Ordering cost which follows the learning effect. |
| $P$ | Selling price per unit. |
| $\theta$ | Decaying rate per unit time. |
| $C$ | Unit purchase cost. |
| $h$ | Unit holding cost which follows the learning effect. |
| $Q$ | Order quantity. |
| $M$ | The offered trade credit by the supplier to the retailer to settle the account. |
| $N$ | The credit financing period for customer offered by the retailer. |
| $I_c$ | Interest charged. |
| $I_e$ | Interest gained. |
| $T$ | Cycle length. |
| $q(t)$ | The inventory level in the interval, $0 \leq t \leq T$ . |
| $K_1(T)$ | The whole inventory level under the condition $M \leq T$ . |
| $K_2(T)$ | The whole inventory level under the condition $N \leq T \leq M$. |
| $K_3(T)$ | The whole inventory level under the condition $N \geq T$. |
| $T_1$ | Cycle length under the $M \leq T$ . |
| $T_2$ | Cycle length under the case $N \leq T \leq M$. |
| $T_3$ | Cycle length under the $N \geq T$. |

## 4.3 Mathematical Formulation

Suppose that, $q(t)$ is the inventory level in the interval $(0 \leq t \leq T)$. Initially, the stock level is $Q$. The present stock is reducing due to demand and deterioration and finally, it has finished at $t = T$. The rate of decreasing of inventory stock are given as follows:

$$\frac{dq(t)}{dt} + \theta q(t) = -R, \quad 0 \leq t \leq T \tag{4.1}$$

With initial and boundary conditions

$$q(0) = Q \text{ and } q(T) = 0 \tag{4.2}$$

the solution of Eq. (4.1) is

$$q(t) = \frac{R}{\theta}\left(e^{\theta(T-t)} - 1\right), \quad 0 \leq t \leq T \tag{4.3}$$

Using the Eq. (4.2), the order quantity given as

$$Q = q(0) = \frac{R}{\theta}\left(e^{\theta T} - 1\right) \tag{4.4}$$

Now, the ordering cost per cycle,

$$OC = \frac{1}{T}\left(C_1 + \frac{C_2}{n^\beta}\right) \tag{4.5}$$

The holding cost per cycle,

$$IHC = \frac{\left(h_1 + \frac{h_2}{n^\beta}\right)R}{\theta^2 T}\left(e^{\theta T} - \theta T - 1\right) \tag{4.6}$$

The deterioration cost per cycle,

$$CD = C(Q - RT) = \frac{CR}{\theta T}(e^{\theta T} - \theta T - 1) \tag{4.7}$$

$$\text{Preservation cost,} \quad PV = \xi T \tag{4.8}$$

Now the whole cost per cycle

$$K(T) = \frac{1}{T}[IHC + OC + CD + PV - IE + IC]$$

Regarding interest charged and interest earned, based on the length of the cycle time $T$, three cases may arise:

**Case-1**: $M \leq T$.

**Case-2**: $N \leq T \leq M$.

**Case-3**: $N \geq T$.

These three cases are represented in Figs. 4.1, 4.2 and 4.3 respectively.

**Case-1**: $M \leq T$ (Fig. 4.1).

During the credit period, the retailer sells items and deposits the generated revenue into an amount bearing account at the interest rate $I_e$ per dollar per year.

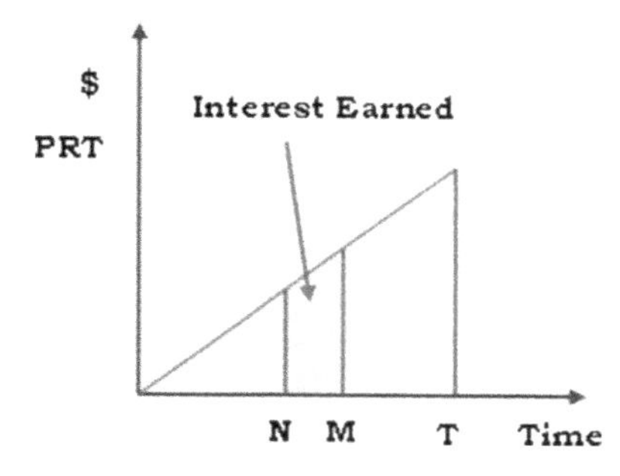


**Fig. 4.1** The total accumulation of interest earned when $M \leq T$ (*Source* own)

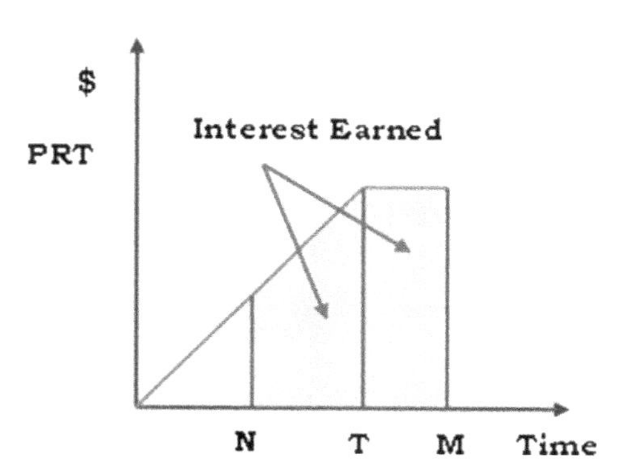


**Fig. 4.2** The total accumulation of interest earned when $N \leq T \leq M$

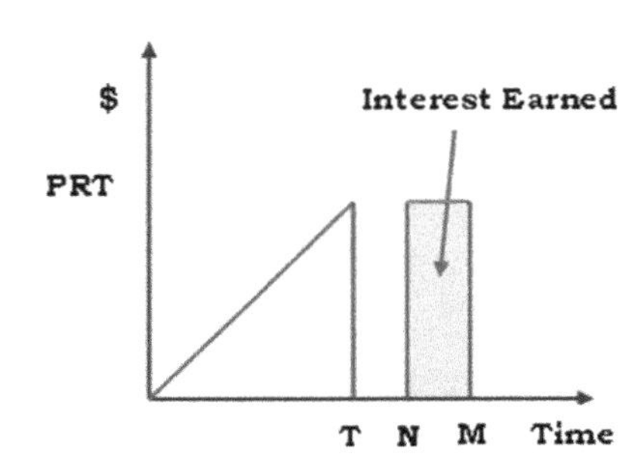


**Fig. 4.3** The total accumulation of interest earned when $N \geq T$ (*Source* own)

Therefore, the interest earned per unit time is

$$IE_1 = \frac{PI_e}{T} \int_N^M Rt\mathrm{d}t = \frac{PI_e R(M^2 - N^2)}{2T} \tag{4.9}$$

The unsold items in stock are charged at interest rate $I_c$ by the supplier at the beginning of time $T$.

Therefore, the interest charged per unit time is

$$IC_1 = \frac{CI_c}{T} \int_M^T q(t)\mathrm{d}t = \frac{CI_c R}{\theta^2 T}\left[e^{\theta(T-M)} - \theta(T-M) - 1\right] \tag{4.10}$$

Hence, the total cost per time unit is

$$K_1(T) = OC + IHC + CD + PV + IC_1 - IE_1 \tag{4.11}$$

**Case 2**: $N \leq T \leq M$ (Fig. 4.2).

In this case, the total interest earned per unit time is,

$$\begin{aligned} IE_2 &= \frac{PI_e}{T}\left[\int_N^T Rt\mathrm{d}t + RT(M - T)\right] \\ &= \frac{PI_e R}{2T}\left(2MT - N^2 - T^2\right) \end{aligned} \tag{4.12}$$

In this case, total interest charged $= 0$

Hence, the total cost per time unit is

$$K_2(T) = OC + IHC + PV + CD - IE_2 \tag{4.13}$$

**Case 3**: $N \geq T$ (Fig. 4.3).

The interest earned per unit time is,

$$IE_3 = PI_eR(M - N) \tag{4.14}$$

Similar as case 2, total interest charged equal to zero.
Hence, the total cost per unit time is,

$$K_3(T) = OC + IHC + PV + CD - IE_3 \tag{4.15}$$

Hence, the total relevant cost $K(T)$ per unit time is,

$$K(T) = \begin{cases} K_1(T), \text{ if } M \leq T \\ K_2(T), \text{ if } N \leq T \leq M \\ K_3(T), \text{ if } N \geq T \end{cases} \tag{4.16}$$

where under series approximation and assumption that $\theta T < 1$, ignoring higher powers of $\theta$, then we get,

$$\begin{aligned} K_1(T) = {} & \frac{\left(C_1 + \frac{C_2}{n^\beta}\right)}{T} + \frac{\left(h_1 + \frac{h_2}{n^\beta}\right)RT}{2} + \frac{CR\theta T}{2} + \frac{CI_cRT}{2} \\ & + \frac{CI_cRM^2}{2T} + \frac{\xi T}{T} - CI_cRM - \frac{PI_eR\left(M^2 - N^2\right)}{2T} \end{aligned} \tag{4.17}$$

and

$$\begin{aligned} K_2(T) = {} & \frac{\left(C_1 + \frac{C_2}{n^\beta}\right)}{T} + \frac{\left(h_1 + \frac{h_2}{n^\beta}\right)RT}{2} + \frac{\xi T}{T} \\ & + \frac{CR\theta T}{2} - \frac{PI_eR}{2T}\left(2MT - N^2 - T^2\right) \end{aligned} \tag{4.18}$$

and

$$K_3(T) = \frac{\left(C_1 + \frac{C_2}{n^\beta}\right)}{T} + \frac{\left(h_1 + \frac{h_2}{n^\beta}\right)RT}{2} + \frac{\xi T}{T} + \frac{CR\theta T}{2} - PI_eR(M - N) \tag{4.19}$$

The necessary and sufficient conditions for $K_1(T)$ to be optimum is

$$\begin{aligned} \frac{\mathrm{d}K_1(T)}{\mathrm{d}T} = {} & -\frac{\left(C_1 + \frac{C_2}{n^\beta}\right)}{T^2} + \frac{\left(h_1 + \frac{h_2}{n^\beta}\right)R}{2} + \frac{CR\theta}{2} \\ & + \frac{CI_cR}{2}\left(1 - \frac{M^2}{T^2}\right) + \frac{PI_eR\left(M^2 - N^2\right)}{2T^2} \end{aligned} \tag{4.20}$$

and

$$\frac{\mathrm{d}^2 K_1(T)}{\mathrm{d}T^2} = \frac{2\left(C_1 + \frac{C_2}{n^\beta}\right)}{T^3} + \frac{CI_c RM^2}{T^3} - \frac{PI_e R\left(M^2 - N^2\right)}{T^3} > 0 \tag{4.21}$$

For the optimal cycle time $T_1$, set $\frac{\mathrm{d}K_1(T)}{\mathrm{d}T} = 0$ which gives

$$T = T_1(\text{say}) = \sqrt{\frac{2\left(C_1 + \frac{C_2}{n^\beta}\right) + RCI_c M^2 + RPI_e\left(N^2 - M^2\right)}{R\left\{h_1 + \frac{h_2}{n^\beta} + C(\theta + I_c)\right\}}} \tag{4.22}$$

Now,

$$\frac{\mathrm{d}K_2(T)}{\mathrm{d}T} = -\frac{C_1 + \frac{C_2}{n^\beta}}{T^2} + \frac{\left(h_1 + \frac{h_2}{n^\beta}\right)R}{2} + \frac{CR\theta}{2} + \frac{PI_e R}{2} - \frac{PI_e RN^2}{2T^2} \tag{4.23}$$

and

$$\frac{\mathrm{d}^2 K_2(T)}{\mathrm{d}T^2} = \frac{2\left(C_1 + \frac{C_2}{n^\beta}\right)}{T^3} + \frac{PI_e RN^2}{T^3} > 0 \tag{4.24}$$

For the optimal cycle time $T_2$, set $\frac{dK_2(T)}{dT} = 0$ which gives

$$T = T_2(\text{say}) = \sqrt{\frac{2\left(C_1 + \frac{C_2}{n^\beta}\right) + PI_e RN^2}{R\left(h_1 + \frac{h_2}{n^\beta} + PI_e + C\theta\right)}} \tag{4.25}$$

Now,

$$\frac{\mathrm{d}K_3(T)}{\mathrm{d}T} = -\frac{C_1 + \frac{C_2}{n^\beta}}{T^2} + \frac{\left(h_1 + \frac{h_2}{n^\beta}\right)R}{2} + \frac{CR\theta}{2} \tag{4.26}$$

and

$$\frac{\mathrm{d}^2 K_3(T)}{\mathrm{d}T^2} = \frac{2\left(C_1 + \frac{C_2}{n^\beta}\right)}{T^3} > 0 \tag{4.27}$$

For the optimal cycle time $T_3$, set $\frac{\mathrm{d}K_3(T)}{\mathrm{d}T} = 0$ which gives,

$$T = T_3(\text{say}) = \sqrt{\frac{2\left(C_1 + \frac{C_2}{n^\beta}\right)}{R\left(h_1 + \frac{h_2}{n^\beta} + C\theta\right)}} \tag{4.28}$$

### 4.3.1 Algorithm

**Step-1**: Compute $T_1$, $T_2$ and $T_3$ from the Eqs. (4.22), (4.25) and (4.28) with the help of input parameters.
**Step-2**: If $M \leq T_1$, then calculate $K_1(T_1)$, otherwise go to step-3.
**Step-3**: If $N \leq T_2 \leq M$, then calculate $K_2(T_2)$, otherwise go to step-4.
**Step-4**: If $N \geq T_3$, then calculate $K_3(T_3)$.
**Step-5**: Find corresponding cycle time and total optimal cost.

### 4.3.2 Numerical Example

$R = 500$ units, $h_1 = 2, h_2 = 1, C_1 = 30, C_2 = 10, \beta = 0.23, \theta = 0.20, \xi = 0.15$ per items, Ie $= \$0.14/\$/$year, $Ip = \$0.15/\$/$year, $C = \$50$, $M = 45/365$ year, $N = 25/365$ year, optimal cycle length, $T^* = 2.7217$ year and minimum total cos $t$
$\psi_1(T^*) = 3253$ \$ per year

### 4.3.3 Sensitive Analysis and Discussion Part

**Sensitive Analysis**
The sensitive analysis has been studied on the effective parameters and retailer total cost will be analyzed.

**Managerial Insights**

- From Table 4.1, if learning rate increases, cycle length almost fixed and retailer's total cost decreases.
- From Table 4.2, if number of shipments increases, cycle length increases marginally and retailer's cost decreases due to the learning effect.
- From Table 4.3, if $M$ increases then cycle length and retailer's total cost decrease due to the learning and credit financing.

**Table 4.1** Impact of learning rate under cycle time and whole cost per cycle (*Source* own)

| Learning rate $\beta$ | Cycle length $T$ (Year) | Retailer's total cost $K_1(T)$ (\$) |
|---|---|---|
| 0.23 | 2.7217 | 3253 |
| 0.24 | 2.7217 | 3247 |
| 0.25 | 2.7217 | 3240 |
| 0.26 | 2.7217 | 3232 |
| 0.27 | 2.7217 | 3228 |

**Table 4.2** Impact of the number of shipments on cycle time and whole cost (*Source* own)

| Number of shipments ($n$) | Cycle length $T$(year) | Retailer's total cost $K_1(T)$ ($) |
|---|---|---|
| 1 | 2.7215 | 3433 |
| 2 | 2.7216 | 3347 |
| 3 | 2.7216 | 3303 |
| 4 | 2.7217 | 3274 |
| 5 | 2.7217 | 3253 |

**Table 4.3** Impact of the number of credit period on cycle time and whole cost (*Source* own)

| Credit period $M$ (year) | Cycle time $T$(Year) | Retailer's total cost $K_1(T)$ ($) |
|---|---|---|
| 30/365 | 2.7564 | 3274 |
| 35/365 | 2.7550 | 3267 |
| 40/365 | 2.7334 | 3260 |
| 45/365 | 2.7217 | 3253 |
| 50/365 | 2.7098 | 3246 |

- From Table 4.4, if deterioration rate increases, cycle length decreases and retailer's cost decreases.
- From Table 4.5, if customer's credit period increases then, cycle length is fixed and retailer's total cost increases marginally.

**Table 4.4** Impact of the decaying rate on cycle time and whole cost (*Source* own)

| Deterioration rate $\theta$ | Cycle time $T$ (Year) | Retailer's total cost $K_1(T)$ ($) |
|---|---|---|
| 0.10 | 5.5126 | 2597 |
| 0.15 | 3.6497 | 2925 |
| 0.20 | 2.7217 | 3253 |
| 0.25 | 2.1673 | 3581 |
| 0.30 | 1.7991 | 3909 |

**Table 4.5** Impact of the customer's credit period on retailer's cycle time and whole cost (*Source* own)

| Customer's trade credit period $N$ | Cycle length $T$ (year) | Retailer's total cost $K_1(T)$ ($) |
|---|---|---|
| 20/365 | 2.7212 | 3253.30 |
| 24/365 | 2.7217 | 3253.40 |
| 28/365 | 2.7223 | 3253.59 |
| 32/365 | 2.7230 | 3253.76 |

**Table 4.6** Impact of the preservation cost on retailer's cycle time and whole cost (*Source* own)

| Preservation cost $\xi$/items | Cycle length $T$ (year) | Retailer's total cost $K_1(T)$ ($) |
|---|---|---|
| 0.15 | 2.7217 | 3253 |
| 1.15 | 2.7217 | 3254 |
| 2.15 | 2.7217 | 3255 |
| 3.15 | 2.7217 | 3256 |

- From Table 4.6, if preservation cost increases then, cycle length is fixed and retailer's total cost increases.

**Discussion Part**

In this part, we have discussed about the distinct cases and have tried to determine as to which case should be better for this model after we procured the solution with the assistance of the concerned algorithm. Post getting all the values from the above three cases, we concluded that the minimum cost was given by Case 1 which is $N \leq M \leq T$ and it was beneficial due to suitable credit period which have obtained from the algorithm and other cases are not consider due the large value of credit period and have been analyzed from the algorithm. This case provided the optimal values of all the parameters. The learning effect will reduce the total cost with trade-credit policy.

## 4.4 Conclusion

This article developed an inventory model to determine cycle length and the corresponding total cost for the buyer with the help of the two-level trade credit financing with learning effect applied over the holding cost and the ordering cost. Eventually, we have concluded that results of this model showed that the retailer's total cost reduces as learning parameter value increases. When items are perishable then preservation of the items are must to control the deterioration rate, but the total cost increases marginally. This article reveals that the combination of trade-credit financing and learning concept is very beneficial to get more profit in real scenario. Findings together with mathematical analysis clearly recommended that the existence of two-level trade-credit and effect of leaning have positive effects on the total cost. Present work can be extended such as stock depended demand, and cloudy environment etc.

## References

Abad PL, Jaggi CK (2003) A joint approach for setting unit price and the length of the credit period for a seller when end demand is price sensitive. Int J Prod Econ 83(2):115–122
Agarwal R, Mittal M (2019) Inventory classification using multilevel association rule mining. Int J Decis Support Syst Technol 11(2):1–12
Chen LH, Kang FS (2010) Integrated inventory models considering the two-level trade credit policy and a price-negotiation scheme. Eur J Oper Res 205(1):47–58
Goyal SK (1985) Economic order quantity under conditions of permissible delay in payments. J Oper Res Soc 36:35–38
Huang YF (2003) The deterministic inventory models with shortage and defective items derived without derivatives. J Stat Manag Syst 6(2):171–180
Huang YF (2006) An inventory model under two levels of trade credit and limited storage space derived without derivatives. Appl Math Model 30(5):418–436
Huang YF (2007) Optimal retailer's replenishment decisions in the EPQ model under two levels of trade credit policy. Eur J Oper Res 176(3):1577–1591
Huang YF, Chung KJ (2003) Optimal replenishment and payment policies in the EOQ model under cash discount and trade credit. Asia Pac J Oper Res 20(2):177–190
Huang YF, Hsu KH (2008) An EOQ model under retailer partial trade credit policy in supply chain. Int J Prod Econ 112(2):655–664
Jaber MY, Goyal SK, Imran M (2008) Economic production quantity model for items with imperfect quality subjected to learning effects. Int J Prod Econ 115:143–150
Jaggi CK, Goyal SK, Goel SK (2008) Retailer's optimal replenishment decisions with credit-linked demand under permissible delay in payments. Eur J Oper Res 190(1):130–135
Jaggi CK, Khanna A, Mittal M (2011) Credit financing for deteriorating imperfect-quality items under inflationary conditions. Int J Serv Oper Inf 6(4):292–309
Jayaswal M, Sangal I, Mittal M, Malik S (2019) Effects of learning on retailer ordering policy for imperfect quality items with trade credit financing. Uncertain Supply Chain Manag 7:49–62
Jayaswal MK, Mittal M, Sangal I (2021) Ordering policies for deteriorating imperfect quality items with trade-credit financing under learning effect. Int J Syst Assur Eng Manag 12(1):112–125
Luo J (2007) Buyer–vendor inventory coordination with credit period incentives. Int J Prod Econ 108(1–2):143–152
Mittal M, Khanna A, Jaggi CK (2017) Retailer's ordering policy for deteriorating imperfect quality items when demand and price are time-dependent under inflationary conditions and permissible delay in payments. Int J Procurement Manage 10(4):461–494
Sarmah SP, Acharya D, Goyal SK (2007) Coordination and profit sharing between a manufacturer and a buyer with target profit under credit option. Eur J Oper Res 182(3):1469–2147
Shah NH, Gor AS, Wee HM (2010) An integrated approach for optimal unit price and credit period for deteriorating inventory system when the buyer's demand is price sensitive. Am J Math Manag Sci 30(3–4):317
Shah N, Pareek S, Sangal I (2012) EOQ in fuzzy environment and trade credit. In J Ind Eng Computations 3(2):133–144.330
Shah NH, Pareek S, Sangal I (2016) Deteriorating inventory model under permissible delay in payments and fuzzy environment. In: Optimal Inventory Control Manag Techn IGI Global.
Shinn SW, Hwang H (2003) Optimal pricing and ordering policies for retailers under order-size-dependent delay in payments. Comput Oper Res 30:35–50
Su CH, Ouyang LY, Ho CH, Chang CT (2007) Retailer's inventory policy and supplier's delivery policy under two-level trade credit strategy. Asia-Pacific J Oper Res 24(05):613–630
Teng JT, Chang CT (2009) Optimal manufacturer's replenishment policies in the EPQ model under two levels of trade credit policy. Eur J Oper Res 195(2):358–363
Teng JT, Goyal SK (2007) Optimal ordering policies for a retailer in a supply chain with up-stream and down-stream trade credits. J Opera Res Soc 58(9):1252–1255
Wright TP (1936) Factors affecting the cost of airplanes. J Aeronaut Sci 3:122–128

# Chapter 5
# An Inventory Policy for Maximum Fixed Life-Time Item with Back Ordering and Variable Demand Under Two Levels Order Linked Trade Credits

**Mrudul Y. Jani, Nita H. Shah, and Urmila Chaudhari**

**Abstract** In this chapter, an inventory policy of the item with maximum fixed life-time is studied where two levels of trade credit depend on the order quantity. We consider the inventory system in which the supplier is ready to give a mutually agreed credit period to the retailer only if the order quantity purchased by the retailer is larger than the predetermined order quantity. Moreover, to be more practical, the retailer offers a credit limit to the customers. Here, price and time-sensitive demand are debated under the inflationary environment over the finite time horizon. In this study, the shortage is allowed and it is fully backordered. The main objective is to maximize the total profit of the retailer to the fraction of the replenishment cycle and the number of replenishments during the planning horizon. The model is supported by numerical examples. Sensitivity analysis is carried out to derive insights for decision-makers.

**Keywords** Inventory · Order linked trade credit · Inflation with time value of money · Maximum fixed life-time · Price-sensitive demand · Shortage

## 5.1 Introduction

In traditional business transactions, it was assumed that the buyer must pay the procurement cost when the products are received. However, in today's competitive markets most companies offer buyer various credit terms like permissible delay in payment, cash discount, etc. to simulate sales and hence reduce inventory. Trade credit has been widely used to boost sales and reduce default risk and attract new customers. In a review of literature for inventory models with trade credit funding,

M. Y. Jani (✉)
Department of Applied Sciences, Faculty of Engineering and Technology, Parul University, Vadodara, Gujarat 391760, India

N. H. Shah
Department of Mathematics, Gujarat University, Ahmedabad, Gujarat 380009, India

U. Chaudhari
Government Polytechnic Dahod, Dahod, Gujarat 389151, India



N. H. Shah et al. (eds.), *Decision Making in Inventory Management*,
Inventory Optimization, https://doi.org/10.1007/978-981-16-1729-4_5

Goyal (1985) was the first researcher who studied the effect of trade credit on optimal inventory policies. Thereafter, based on the work of Goyal (1985) many researchers developed inventory models under trade credit. For example, Shah (1993) considered a stochastic inventory model for deteriorating item when a delay in payments are permissible. Later, Aggarwal and Jaggi (1995) extended the EOQ model from non-deteriorating items to deteriorating items. Jamal et al. (1997) further generalized the EOQ model with trade credit to allow shortages. Then, Huang (2003) extended the trade credit problem where the up-stream credit limit is greater than down-stream trade credit. Also, Liao (2008) extended Huang's model to an economic production quantity model for deteriorating items. Subsequently, Ouyang et al. (2013) proposed an EOQ model in which they relaxed two assumptions of Huang's (2003) work. Chen et al. (2014) proposed an EOQ model under different credit terms. Shah and Jani (2016a) expressed optimal ordering policies for two-level trade credit with quadratic demand. Shah et al. (2018) built optimal ordering policies for deteriorating items with permissible delay in payment options. Recently, Jani et al. (2020) determined inventory control policies for deteriorating items under the two-level order linked trade credit. Other articles related to this area are Rabbani et al. (2018), Mahata and Mahata (2020), and others.

In inventory Modelling, the demand rate was taken as a constant but the common characteristic of the articles in traditional inventory assumption of a constant demand rate is generally valid in the mature stage of the total life of the product. Though, it is rare in real life with this limitation Pal et al. (1993) first considered the inventory model with stock-dependent demand. However, in real-life situations, the demand may increase or fall with time and price also. Considering this Wee (1997) developed an inventory model for price-sensitive demand. Wu (2001) further investigated the inventory model by considering the ramp type demand rate and Weibull distribution deterioration. Jaggi et al. (2008) presented the retailer's optimal replenishment strategies for the permissible delay in payment dependent demand. Shah et al. (2015) developed an EOQ model for price-sensitive quadratic demand. Later, Shah et al. (2017a) studied the retailer's optimum policies for price-credit dependent trapezoidal demand. Recently, Chaudhari et al. (2020) investigated the inventory model taking account of deteriorating effects, preservation, advanced payment scheme under quadratic demand. Other articles related to this area are Bose et al. (1995), Shah et al (2016), and others.

In many real-life situations, products deteriorate continuously such as volatile liquids, medicines, blood banks, drugs, food, and others when kept in storage for a long period. For such products, losses due to deterioration cannot be ignored. Inventory problems related to deteriorating items have been studied widely in earlier research. In this regard, Ghare and Schrader (1963) proposed an EOQ model for an exponentially decaying item. Philip (1974) then obtained an inventory model with a three-parameter Weibull distribution deterioration rate. Later, Raafat (1991) provided a survey of literature on continuously deteriorating inventory policies. Sett et al. (2012) analyzed a two-warehouse inventory model with time-dependent deterioration. Sarkar et al. (2015) discussed an inventory control policy with a maximum fixed lifetime of the product. Shah and Jani (2016b) studied an EOQ model for non-instantaneous deterioration items under order size dependent trade credit for price-dependent demand. Lately, Shah et al. (2017b) built an inventory model under three different cases, i.e., the maximum fixed life of the product, constant deterioration rate, and without deterioration rate. Recently, Gautam et al. (2020) developed an inventory system by considering the detrimental impacts of deterioration.

Most of the inventory models that grew so far do not include inflation and the time value of money as parameters of the system. Maybe low inflation in the economy of the western nations before the 1970s was at the foundation of this methodology in inventory modeling. Nowadays, inflation has become a permanent feature of the economy throughout the world. This changing situation in the world economy didn't get away from the consideration of the inventory modelers. Buzacott (1975) introduced the EOQ model with a uniform inflation rate for all associated costs. Ray and Chaudhuri (1997) and Chang et al. (2009) developed an inventory model with inflation. Sarkar (2012) discussed an inventory model of the finite renewal rate. Lashgari et al. (2016) investigated an inventory control model for back-ordering and credit limit with inflation. Shaikh et al. (2017) proposed an inventory model to consider inflation, fully backlogged shortages with stock-dependent demand. Other papers are related to this area are Shah et al. (2017c), Gupta et al. (2020), and others.

In this chapter, we develop an inventory model over a finite planning horizon, with maximum fixed lifetime deterioration, price and time-varying demand rate, inflation, time value of money order quantity linked to delay in payment, and shortage and fully backordering. The proposed inventory model represents one supplier-one retailer for a two-layer of trade credit options where the supplier gives credit period to the retailer only if the stock purchased by the retailer is more than the pre-scheduled order quantity. The main objective is to maximize the net current value of the retailer's profit. In Sect. 5.2, the notation and assumptions are recognized. Section 5.3, includes the formulation of the mathematical model. Section 5.4 gives a sensitivity analysis of the optimal result, it proved by the help of some numerical examples. In Sect. 5.5, the conclusion and future work have been given.

## 5.2 Notations and Assumptions

### *5.2.1 Notations*

| | |
|---|---|
| $A$ | Setup cost ($/lot). |
| $h$ | Holding cost/unit/unit time (in $). |
| $k$ | The net rate of constant decline in inflation. |
| $C$ | Purchasing cost ($/unit) at time . |
| $C(t)$ | $Ce^{-kt}$; The unit purchasing cost at any time $t$ ($/unit). |
| $p$ | Selling price ($/unit) at time $t = 0$; $p > C$. |
| $p(t)$ | $pe^{-kt}$; Selling price ($/unit) at time $t$. |
| $M$ | Supplier deals with permissible delay in payment to the retailer (in years). |

| | |
|---|---|
| $N$ | Retailer deals with permissible delay in payment to the customer (in years). |
| $H$ | Finite planning horizon (in years). |
| $\delta$ | Shortage cost (/\$/unit/timeunit). |
| $I_c$ | The interest rate paid by the retailer to the supplier (/\$/year); $I_c > I_e$. |
| $I_e$ | Rate of interest earned by the retailer (/\$/year). |
| $I_b$ | The interest rate paid by the retailer to the bank (/\$/year) if $T > M$. |
| $n$ | Replenishments' number (Decision Variable). |
| $F$ | The fraction of the replenishment cycle, in which inventory level is positive (in years) (Decision Variable). |
| $T$ | $\frac{H}{n}$; Replenishment or cycle time (in years). |
| $t_1$ | $\frac{FH}{n}$; the time that inventory level reaches to zero in the $j$th replenishment (in years). |
| $m$ | Item's maximum fixed life (in years). |
| $Q$ | Order quantity in each replenishment (in units). |
| $Q_d$ | Predetermined order quantity (in units). |
| $I(t)$ | Inventory's level at any point in time $t$ (in units). |
| $I_m$ | The maximum inventory level (in units). |
| $I_s$ | The maximum shortage quantity (in units). |
| $\pi(F, n)$ | Total profit of retailer for $n$th replenishment (in \$). |

### 5.2.2 Assumptions

1. The planning horizon is finite.
2. The inventory policy deals with a single product.
3. The shortage is allowed and it is completely backordered.
4. Demand rate, (say) $R(p, t) = a - bp(t)$; where $a > 0$ is scale demand and $b > 0$ is a mark-up of the selling price.
5. $\theta(t) = \frac{1}{1+m-t}$, $0 \leq t \leq T \leq m$ is instantaneous deterioration. Where $\theta(t) \leq 1$ for any time $m$.
6. No interest paid or received for the quantity of stuff which has been deteriorated in the period of $\frac{FH}{n}$.
7. The supplier is ready to give a mutually agreed credit period $M$ to the retailer only if stock purchased by the retailer is larger than the prearranged order quantity. i.e. $Q > Q_d$.
8. The retailer pays interest rates $I_b$ to the bank for $T > M$.
9. The constant inflation rate is considered with the time value of money.
10. Lead time is zero or negligible.

## 5.3 Mathematical Model

Under the assumptions, the rate of inventory level $I(t)$ at the time $t$ during the period $[0, t_1]$ is governed by the following differential equation (Fig. 5.1)

$$\frac{\mathrm{dI_1(t)}}{\mathrm{dt}} + \theta(t)I_1(t) = -R(p, t), \quad 0 \le t \le t_1, \tag{5.1}$$

with the boundary condition $I_1(0) = I_m$ and $I_1(t_1) = 0$.

$$\frac{\mathrm{dI_2(t)}}{\mathrm{dt}} = -R(p, t), \quad t_1 \le t \le T, \tag{5.2}$$

with the boundary condition $I_2(t_1) = 0$ and $I_2(T) = -I_s$.

Using boundary condition $I_1(t_1) = 0$ from (5.1), one has

$$\begin{aligned} I_1(t) &= (1 + m - t)W_1 1\left(n\left(\frac{1 + m - t}{1 + m - t_1}\right) + W_2(t - t_1)\right. \\ &\left. - \frac{bpk^2}{4}\left(t^2 - t_1^2\right)\right), \quad 0 \le t \le t_1 \\ &\text{where} \quad W_1 = a - bp + bpk(1 + m) \\ &- \frac{bpk^2(1 + m)^2}{2} \text{and } W_2 = bpk - \frac{bpk^2(1 + m)}{2} \end{aligned} \tag{5.3}$$

Now, with the help of the boundary condition $I_1(0) = I_m$, the maximum inventory level is

$$I_m = (1 + m)\left(W_1 \ln\left(\frac{1 + m}{1 + m - t_1}\right) - W_2 t_1 + \frac{1}{4}bpk^2 t_1^2\right) \tag{5.4}$$

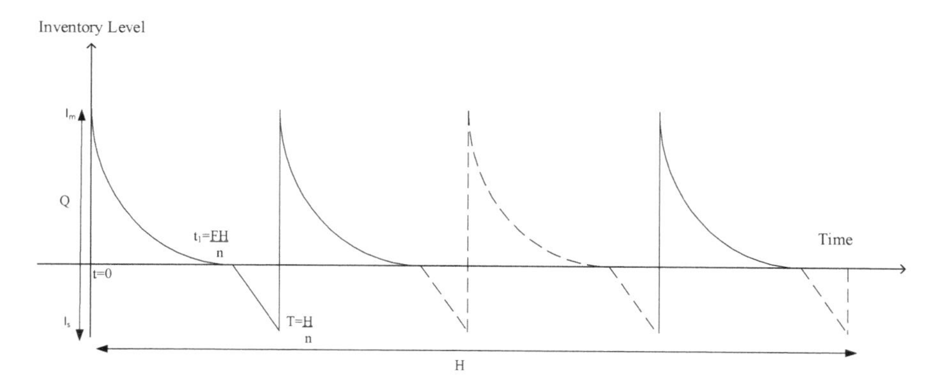


**Fig. 5.1** Graphical representation of the inventory system Lashgari et al. (2016)

From (5.2) with $I_2(t_1) = 0$, anyone has

$$I_2(t) = a(t_1 - t) + bp(t - t_1 + \frac{k}{2}(t_1^2 - t^2) + \frac{k^2}{6}(t^3 - t_1^3)),\ t_1 \le t \le T \tag{5.5}$$

With the boundary condition $I_2(T) = -I_s$, the maximum shortage level is

$$I_s = -a\,(t_1 - T) - bp\left(T - t_1 + \frac{1}{2}k(t_1^2 - T^2) + \frac{1}{6}k^2(T^3 - t_1^3)\right) \tag{5.6}$$

Thus, the total ordering quantity is

$$\begin{aligned} Q =& I_m + I_s \\ =& (1+m)\left(W_1 \ln\left(\frac{1+m}{1+m-t_1}\right) - W_2 t_1 + \frac{1}{4}bpk^2t_1^2\right) \\ & - a\,(t_1 - T) - bp\left(T - t_1 + \frac{1}{2}k(t_1^2 - T^2) + \frac{1}{6}k^2(T^3 - t_1^3)\right) \end{aligned} \tag{5.7}$$

Relevant costs of the retailer's total profit are as follow:

- Ordering/Setup Cost: $OC = \sum_{j=0}^{n} Ae^{-jkT}$
- Holding Cost: $HC = Ch\left[\sum_{j=0}^{n-1} e^{-jkT} \int_0^{t_1} I_1(t)\mathrm{d}t\right]$
- Purchasing Cost: $PC = CQ\left[\sum_{j=0}^{n-1} e^{-jkT}\right]$
- Sales Revenue: $SR = \left[\sum_{j=0}^{n-1} e^{-jkT} \int_0^{T} p(t)R(p,t)\mathrm{d}t\right]$
- Shortage cost: $SC = \delta\left[\sum_{j=0}^{n-1} e^{-jkT} \int_{t_1}^{T} -I_2(t)\mathrm{d}t\right]$

Since we study two levels of trade credit policies linked to order quantity. The supplier proposes a credit limit $M$ to a retailer if the order size is more than $Q_d$, otherwise at the time of receiving an order, the retailer pays the total purchasing cost. However, in both cases, the retailer gives trade credit $N$ to customers. Therefore, from the above conversation, two possible cases will arise (1) $Q < Q_d$ and (2) $Q > Q_d$.

**Case 1:** $Q < Q_d$.

In these circumstances, the predetermined order quantity $Q_d$ is more than the retailer's order quantity $Q$. So, the supplier does not offer trade credit $M$ to a retailer. However, the retailer offers trade credit $N$ to the end customers. As an outcome retailer takes a loan from a bank at time zero and $N$ starts to pay back. Thus, interest payable to suppliers and banks by the retailer are $IC_1 = CI_c \sum_{j=0}^{n-1} e^{-jkT}\left(\int_0^{t_1} R(p,t)t\,\mathrm{d}t + \int_0^{N} R(p,t)T\,\mathrm{d}t\right)$, and $CC_1 = CI_b(\sum_{j=0}^{n-1} e^{-jkT} \int_0^{t_1} I_1(t)\mathrm{d}t)$ respectively.

Hence, the total profit all over the finite planning horizon is

$$\pi_1(F, n) = SR - HC - SC - OC - PC - IC_1 - CC_1 \tag{5.8}$$

**Case 2:** $Q > Q_d$.

In this situation, the order quantity $Q$ of the retailer is more than an order quantity $Q_d$. Therefore, the supplier gives trade credit $M(M \neq 0)$ to a retailer. Thus, possible subcases are as follows:

$$\begin{cases} \text{Subcase-1: } M < N \; M < t_1 \\ \text{Subcase-2: } M < N \; t_1 < M \\ \text{Subcase-3: } M > N \; M < t_1 \\ \text{Subcase-4: } M > N \; t_1 < M < t_1 + N \\ \text{Subcase-5: } M > N \; t_1 + N < M \end{cases}$$

*Subcase-1*: $M < N$ and $M < t_1$. In this subcase, due to $M < N$, interest earned to the retailer is zero. On the other hand, interest payable by the retailer to supplier and bank are $IC_2 = CI_c \sum_{j=0}^{n-1} e^{-jkT} (\int_0^{t_1} tR(p,t)dt + \int_0^{N-M} TR(p,t)\mathrm{d}t)$, and $CC_2 = CI_b(\sum_{j=0}^{n-1} e^{-jkT} \int_M^{t_1} I_1(t)\mathrm{d}t)$ respectively.

Consequently, the existing total profit for the duration of the finite planning horizon is

$$\pi_2(F, n) = SR - HC - SC - OC - PC - IC_2 - CC_2 \tag{5.9}$$

*Subcase-2*: $M < N$ and $t_1 < M$. Due to $t_1 < M$, interest payable by the retailer to the bank is zero. Furthermore, since $M < N$, interest earned to the retailer is zero. So, interest payable by the retailer to the supplier is $IC_3 = CI_c \sum_{j=0}^{n-1} e^{-jkT} (\int_0^{N-M} R(p,t)T\,\mathrm{d}t + \int_0^{t_1} tR(p,t)\mathrm{d}t)$.

As a result, the current total profit for the duration of the finite planning horizon is

$$\pi_3(F, n) = SR - HC - SC - OC - PC - IC_3 \tag{5.10}$$

*Subcase-3*: $M > N$ and $M < t_1$. In this subcase, the retailer earns interest by selling the items and he has to pay interest to the supplier and bank as follows:

$$IE_4 = pI_e \sum_{j=0}^{n-1} e^{-jkT} \left( \int_N^M R(p,t)t\,\mathrm{d}t + I_s(M-N) \right),$$

$$IC_4 = CI_c \sum_{j=0}^{n-1} e^{-jkT} \left( \int_0^{t_1+N-M} R(p,t)t\,\mathrm{d}t \right)$$

and

$$CC_4 = CI_b \sum_{j=0}^{n-1} e^{-jkT} \left( \int_M^{t_1} I_1(t) \mathrm{d}t \right)$$

Thus, the existing total profit for the duration of the finite planning horizon is

$$\pi_4(F, n) = SR - HC - OC - PC - IC_4 - CC_4 + IE_4 \tag{5.11}$$

*Subcase-4*: $M > N$ and $t_1 < M < t_1 + N$. Since $t_1 < M$ interest payable by the retailer to the bank is zero. Furthermore, in this situation, the retailer's interest earns and the retailer's interest charge to the supplier are $IE_5 = pI_e \sum_{j=0}^{n-1} e^{-jkT} (\int_N^M R(p, t) t \, \mathrm{d}t + I_m(M - N))$, and $IC_5 = CI_c \sum_{j=0}^{n-1} e^{-jkT} (\int_0^{t_1+N-M} R(p, t) t \, \mathrm{d}t)$ respectively.

So, the current total profit for the duration of the finite planning horizon is

$$\pi_5(F, n) = SR - HC - SC - OC - PC - IC_5 + IE_5 \tag{5.12}$$

*Subcase-5*: $M > N$ and $t_1 + N < M$. In the current subcase, since $t_1 + N < M$:

$$IE_6 = pI_e \sum_{j=0}^{n-1} e^{-jkT} \left( \int_0^{M-N-t_1} R(p, t) T \, \mathrm{d}t + I_s t_1 + \int_0^{t_1} R(p, t) t \, \mathrm{d}t \right)$$

Consequently, the existing total profit for the duration of the finite planning horizon is

$$\pi_6(F, n) = SR - HC - SC - OC - PC + IE_6 \tag{5.13}$$

Hence, total profit is given by

$$\pi(F, n) = \begin{cases} Q < Q_d & \pi_1(F, n) \\ Q > Q_d \begin{cases} M < N, & \begin{cases} \pi_2(F, n), & M < t_1 \\ \pi_3(F, n), & t_1 < M \end{cases} \\ M > N, & \begin{cases} \pi_4(F, n), & M < t_1 \\ \pi_5(F, n), & t_1 < M < t_1 + N \\ \pi_6(F, n), & t_1 + N < M \end{cases} \end{cases} \end{cases}$$

Here, the model considers an algorithm for the optimum solution as follows.

*Algorithm*

**Step 1**: Evaluate $\frac{\partial \pi_i(F,n)}{\partial F}$; $\forall i = 1 \ldots 6$ using the Eqs. (5.8) to (5.13).

**Step 2**: Derive the optimum value of $F$ say $(F_i)$, by calculating $\frac{\partial \pi_i(F,n)}{\partial F} = 0$; $\forall i = 1 \ldots 6$.

**Step 3**: Substitute $F = F_i$, in $\pi_i(F, n)$; $\forall i = 1...6$ where $n \in \mathbb{N}$, using the Eqs. (5.8) to (5.13) respectively.

**Step 4**: Take $n = n + 1$ where $n \in \mathbb{N}$ and once again calculated $\pi_i(F_i, n)$; $\forall i = 1...6$ where $n \in \mathbb{N}$, using the Eqs. (5.8) to (5.13) respectively.

**Step 5**: If $\pi_i(F_i, n+1) < \pi_i(F_i, n)$ ; $\forall i = 1...6$ where $n \in \mathbb{N}$, then the optimal number of replenishments is $n^* = n$ and the corresponding optimal fraction of the replenishment cycle is $F^* = F_i$ otherwise go to step 2.

**Step 6**: Calculate the optimal cycle time $T^* = \frac{H}{n^*}$.

**Step 7**: Calculate $t_1^* = \frac{F^* H}{n^*}$.

**Step 8**: Calculate $I_m^*$ and $I_s^*$.

**Step 9**: Calculate the total optimal profit $\pi_i(F^*, n^*)$; $\forall i = 1...6$.

**Step 10**: Calculate optimum order quantity $Q^* = I_m^* + I_s^*$ from Eq. (5.7).

## 5.4 Numerical Examples with Sensitivity Analysis

### *5.4.1 Numerical Examples*

#### 5.4.1.1 *Numerical Example 1 (M > N)*

Take $a = 1000$ units, $b = 20\%$, $A = \$500$ per order, $C = \$5$ per unit, $I_e = 8\%/\$/\text{year}$, $h = \$0.8$ per unit per unit time, $I_c = 12\%/\$/\text{year}$, $p = \$10$ per unit, $I_b = 8\backslash\%/\$/\text{year}$, $M = 0.15$ year, $k = 0.08$, $N = 0.06$ year, $\delta = 4\,\$/\text{unit}$, $H = 5$ years, and $Q_d = 500$ units.

#### 5.4.1.2 *Numerical Example 2 (M < N)*

Set $N = 0.2$ year in Example 4.1.1.

By solving the numerical examples using the mathematical software Maple XVIII, we have Table 5.1 of optimal solutions.

From Table 5.1, we analyze that in the case ($Q > Q_d$, $M > N$, $M < t_1$), the retailer's total profit is \$13,637.86 which is the maximum, optimal number of replenishments is $n^* = 8$, the optimal fraction of the replenishment cycle is $F^* = 0.249$ year, optimal cycle time is $T^* = 0.625$ year, and an optimum order quantity is $Q^* = 636.21$ units.

The concavity of the retailer's total profit $\pi_4(F^*, n^*)$ for the best optimal case is depicted in Fig. 5.2.

**Table 5.1** Optimal solutions

| Cases | Sub-cases | | Example numbers | Total profit ($) | Decision variables |
|---|---|---|---|---|---|
| $Q < Q_d$ | ** | ** | 1 | Infeasible solution | Infeasible solution |
| $Q > Q_d$ | $M < N$ | $M < t_1$ | 2 | 13,192.15 | $n^* = 8$<br>$F^* = 0.25$ year<br>$Q^* = 636.40$ units<br>$T^* = 0.625$ year |
| | | $t_1 < M$ | 2 | Infeasible solution | Infeasible solution |
| | $M > N$ | $M < t_1$ | 1 | **13,637.86** | $n^* = 8$<br>$F^* = 0.249$ year<br>$Q^* = 636.21$ units<br>$T^* = 0.625$ year |
| | | $t_1 < M < t_1 + N$ | 1 | Infeasible solution | Infeasible solution |
| | | $t_1 + N < M$ | 1 | Infeasible solution | Infeasible solution |

***N.B*** ** indicates not applicable case

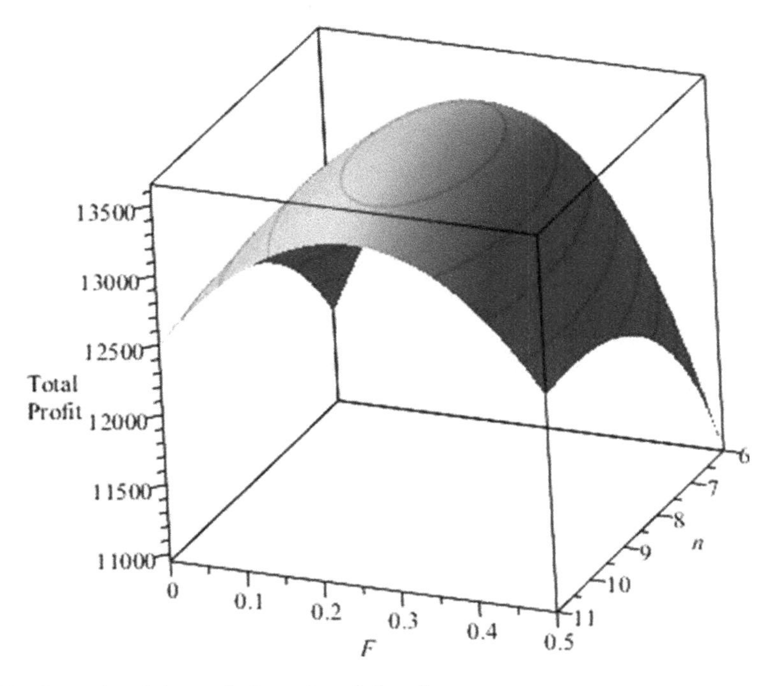


**Fig. 5.2** Concavity of the retailer's total profit function

### 5.4.2 *Sensitivity Analysis*

Sensitivity analysis of the objective function for Example 4.1.1 for various inventory parameters is calculated in Table 5.2.

From Table 5.2, the following results are derived: It is visible that in all cases as the scale demand $a$, rate of interest earned $I_e$, finite planning horizon $H$, maximum fixed lifetime of the item $m$, selling price $p$, and length of the credit period $M$ has a huge positive impact on total profit $\pi_4(F^*, n^*)$. On the other hand by increasing $M$, this suggests that permissible delay in payment facilitates the business for both the supplier and retailer. Since the retailer places a bigger order to the supplier which is favorable for his business finally results in higher profit. Conversely, holding cost $h$, purchasing cost $C$, setup cost $A$, a mark-up of the selling price $b$, interest rate $I_c$, net rate of inflation $k$, retailer deals permissible delay in payment to a customer $N$, the interest rate paid by the retailer to the bank $I_b$, and shortage cost $\delta$ decreases total profit $\pi(F, n)$.

## 5.5 Conclusion

This chapter analyses an inventory system with two decision variables, the fraction of the replenishment cycle $F$ and the replenishment number $n$. Supplier offers an order quantity dependent credit limit to the retailer, and the retailer offers an unconditional credit limit to the customers. Also, inflation and time value of money have been considered. In this model, the shortage is allowed which is fully backordered. We develop an algorithm to maximize the retailer's total profit evaluating the optimal order quantity, shortage amount, and a number of replenishments in a finite planning horizon. This research work suggested to manager an enterprise on how to determine the optimal order quantity and number of replenishments in various types of delay periods under the inflation and time value of money. Current research has several possible extensions like the model can be further generalized by stochastic demand and preservation technology investment, quantity discounts, and take more items at a time. Also, it can be further studied for probabilistic demand, fuzzy stochastic demand, etc.

**Table 5.2** Sensitivity analysis

| Inventory parameters | Values | $n^*$ | $F^*$ (year) | Total profit $\pi_4(F^*, n^*)$ ($) |
|---|---|---|---|---|
| $A$ | 500 | 8 | 0.249 | 13,502.26 |
| | 550 | 8 | 0.249 | 13,130.75 |
| | 600 | 8 | 0.249 | 12,759.25 |
| $C$ | 4 | 8 | 0.290 | 17,718.01 |
| | 4.5 | 8 | 0.268 | 15,601.63 |
| | 5 | 8 | 0.249 | 13,502.26 |
| $a$ | 800 | 8 | 0.249 | 10,024.89 |
| | 900 | 8 | 0.249 | 11,763.58 |
| | 1000 | 8 | 0.249 | 13,502.26 |
| $b$ | 0.8 | 8 | 0.249 | 13,536.16 |
| | 0.9 | 8 | 0.249 | 13,519.21 |
| | 1 | 8 | 0.249 | 13,502.26 |
| | 1.1 | 8 | 0.249 | 13,485.31 |
| | 1.2 | 8 | 0.249 | 13,468.36 |
| $h$ | 0.64 | 8 | 0.263 | 13,574.81 |
| | 0.72 | 8 | 0.256 | 13,537.50 |
| | 0.8 | 8 | 0.249 | 13,502.26 |
| | 0.88 | 8 | 0.242 | 13,468.93 |
| $H$ | 5 | 8 | 0.249 | 13,502.26 |
| | 5.5 | 9 | 0.249 | 14,626.16 |
| | 6 | 10 | 0.250 | 15,702.73 |
| $I_c$ | 0.096 | 8 | 0.250 | 13,504.01 |
| | 0.108 | 8 | 0.249 | 13,503.13 |
| | 0.12 | 8 | 0.249 | 13,502.26 |
| | 0.132 | 8 | 0.24 | 13,501.39 |
| | 0.144 | 8 | 0.248 | 13,500.53 |
| $I_e$ | 0.064 | 8 | 0.250 | 13,446.96 |
| | 0.072 | 8 | 0.250 | 13,474.60 |
| | 0.08 | 8 | 0.249 | 13,502.26 |
| | 0.088 | 8 | 0.248 | 13,529.94 |
| | 0.096 | 8 | 0.247 | 13,557.65 |
| $k$ | 0.064 | 8 | 0.254 | 13,987.64 |
| | 0.072 | 8 | 0.251 | 13,741.89 |
| | 0.08 | 8 | 0.249 | 13,502.26 |
| | 0.088 | 8 | 0.246 | 13,268.57 |
| | 0.096 | 8 | 0.244 | 13,040.66 |

(continued)

**Table 5.2** (continued)

| Inventory parameters | Values | $n^*$ | $F^*$ (year) | Total profit $\pi_4(F^*, n^*)$ ($) |
|---|---|---|---|---|
| $m$ | 0.064 | 8 | 0.247 | 13,495.05 |
| | 0.072 | 8 | 0.248 | 13,498.67 |
| | 0.08 | 8 | 0.249 | 13,502.26 |
| | 0.088 | 8 | 0.250 | 13,505.82 |
| | 0.096 | 8 | 0.250 | 13,509.34 |
| $M$ | 0.12 | 8 | 0.248 | 13,393.77 |
| | 0.135 | 8 | 0.249 | 13,448.17 |
| | 0.15 | 8 | 0.249 | 13,502.26 |
| | 0.165 | 8 | 0.25 | 13,556.05 |
| | 0.18 | 8 | 0.25 | 13,609.54 |
| $N$ | 0.048 | 8 | 0.249 | 13,538.77 |
| | 0.054 | 8 | 0.249 | 13,520.68 |
| | 0.060 | 8 | 0.249 | 13,502.26 |
| | 0.066 | 8 | 0.249 | 13,483.50 |
| | 0.072 | 8 | 0.25 | 13,464.41 |
| $I_b$ | 0.064 | 8 | 0.249 | 13,502.27 |
| | 0.072 | 8 | 0.249 | 13,50 2.26 |
| | 0.08 | 8 | 0.249 | 13,502.26 |
| | 0.088 | 8 | 0.249 | 13,502.26 |
| | 0.096 | 8 | 0.249 | 13,502.25 |
| $\delta$ | 4.0 | 8 | 0.249 | 13,502.26 |
| | 4.4 | 9 | 0.270 | 13,245.39 |
| | 4.8 | 9 | 0.288 | 13,004.55 |
| $p$ | 8 | 8 | 0.251 | 5303.03 |
| | 9 | 8 | 0.250 | 9406.68 |
| | 10 | 8 | 0.249 | 13,502.26 |
| | 11 | 8 | 0.249 | 17,589.76 |
| | 12 | 8 | 0.248 | 21,669.19 |

# References


Aggarwal SP, Jaggi CK (1995) Ordering policies of deteriorating items under permissible delay in payments. J Oper Res Soc 46(5):658–662

Bose S, Goswami A, Chaudhari KS (1995) An EOQ model for deteriorating items with linear time-dependent demand rate and shortages under inflation and time discounting. J Oper Res Soc 46(6):771–782

Buzacott JA (1975) Economic order quantities with inflation. J Oper Res Soc 26(3):553–558

Chang CT, Wu SJ, Chen LC (2009) Optimal payment time with deteriorating items under inflation and permissible delay in payment. Int J Syst Sci 40(10):985–993

Chaudhari U, Shah NH, Jani MY (2020) Inventory modelling of deteriorating item and preservation technology with advance payment scheme under quadratic demand. Optim Inventory Manag 69–79

Chen SC, Cárdenas-Barrón LE, Teng JT (2014) Retailer's economic order quantity when the supplier offers conditionally permissible delay in payments link to order quantity. Int J Prod Econ 155(1):284–291

Gautam P, Khanna A, Jaggi CK (2020) Preservation technology investment for an inventory system with variable deterioration rate under expiration dates and price-sensitive demand. Yugoslav J Oper Res 30(3):289–305

Ghare PM, Scharender GH (1963) A model for the exponentially decaying inventory system. J Ind Eng 14(5):238–243

Goyal SK (1985) Economic order quantity under conditions of permissible delay in payments. J Oper Res Soc 36(4):335–338

Gupta RK, Saxena S, Singh V, Singh P, Mishra NK (2020) An inventory ordering model with different defuzzification techniques under inflation. J Comput Theor Nanosci 17(6):2621–2625

Huang YF (2003) Optimal retailers ordering policies in EOQ model under trade credit financing. J Oper Res Soc 54(9):1011–1015

Jaggi CK, Goyal SK, Goel SK (2008) Retailer's optimal replenishment decisions with credit-linked demand under permissible delay in payments. Eur J Oper Res 190(1):130–135

Jamal AM, Sarker BR, Wang S (1997) An ordering policy for deteriorating items with allowable shortage and permissible delay in payment. J Oper Res Soc 48(8):826–833

Jani MY, Shah NH, Chaudhari U (2020) Inventory control policies for time-dependent deteriorating item with variable demand and two-level order linked trade credit. Optim Inventory Manag 55–67

Lashgari M, Taleizadeh AA, Sana SS (2016) An inventory control problem for deteriorating items with back-ordering and financial considerations under two levels of trade credit linked to order quantity. J Ind Manag Optim 12(3):1091–1119

Liao JJ (2008) An EOQ model with noninstantaneous receipt and exponentially deteriorating items under two-level trade credit. Int J Prod Econ 113(2):852–861

Mahata P, Mahata GC (2020) Production and payment policies for an imperfect manufacturing system with discount cash flows analysis in fuzzy random environments. Math Comput Model Dyn Syst 26(4):374–408

Ouyang LY, Yang CT, Chan YL, Cárdenas-Barrón LE (2013) A comprehensive extension of the optimal replenishment decisions under two levels of trade credit policy depending on the order quantity. Appl Math Comput 224(1):268–277

Pal S, Goswami A, Chaudhuri KS (1993) A deterministic inventory model for deteriorating items with stock dependent demand rate. Int J Prod Econ 32(3):291–299

Philip GC (1974) A generalized EOQ model for items with Weibull distribution. AIIE Trans 6(2):159–162

Raafat F (1991) Survey of literature on continuously deteriorating inventory models. J Oper Res Soc 42(1):27–37

Rabbani M, Hejarkhani B, Farrokhi-Asl H, Lashgari M (2018) Optimal credit period and lot size for deteriorating items with expiration dates under two-level trade credit financing and backorder. J Ind Syst Eng 11(4):1–18

Ray J, Chaudhuri KS (1997) An EOQ model with stock-dependent demand, shortage, inflation, and time discounting. Int J Prod Econ 53(2):171–180

Sarkar B (2012) An EOQ model with delay in payments and time varying deterioration rate. Math Comput Model 55(3–4):367–377

Sarkar B, Saren S, Cárdenas-Barrón LE (2015) An inventory model with trade-credit policy and variable deterioration for fixed lifetime products. Ann Oper Res 229(1):677–702

Sett BK, Sarkar B, Goswami A (2012) A two-warehouse inventory model with increasing demand and time varying deterioration. Sci Iranica 19(6):1969–1977

Shah NH (1993) Probabilistic time-scheduling model for an exponentially decaying inventory when delays in payments are permissible. Int J Prod Econ 32(1):77–82

Shah NH, Jani MY (2016a) Optimal ordering for deteriorating items of fixed-life with quadratic demand and two-level trade credit. Optimal Inventory Control Manag Techn 1–16

Shah NH, Jani MY (2016b) Economic order quantity model for non-instantaneously deteriorating items under order-size-dependent trade credit for price-sensitive quadratic demand. AMSE J 37(1):1–19

Shah NH, Jani MY, Chaudhari UB (2016) Impact of future price increase on ordering policies for deteriorating items under quadratic demand. Int J Ind Eng Comput 7(2016):423–436

Shah NH, Jani MY, Chaudhari UB (2017a) Retailer's optimal policies for price-credit dependent trapezoidal demand under two-level trade credit. Int J Oper Quant Manag 23(2):115–130

Shah NH, Jani MY, Chaudhari UB (2017b) Optimal replenishment time for a retailer under partial upstream prepayment and partial downstream overdue payment for quadratic demand. Math Comput Model Dyn Syst 24(1):1–11

Shah NH, Jani MY, Chaudhari UB (2017c) Study of imperfect manufacturing system with preservation technology investment under inflationary environment for quadratic demand: a reverse logistic approach. J Adv Manufac Syst 16(1):1–18

Shah NH, Jani MY, Chaudhari UB (2018) Optimal ordering policy for deteriorating items under down-stream trade credit-dependent quadratic demand with full up-stream trade credit and partial down-stream trade credit. Int J Math Oper Res 12(3):378–396

Shah NH, Jani MY, Shah DB (2015) Economic order quantity model under trade credit and customer returns for price-sensitive quadratic demand. Rev Inv Oper 36(3):240–248

Shaikh AA, Mashud AHM, Uddin MS, Khan MAA (2017) A non-instantaneous deterioration inventory model with price and stock dependent demand for fully backlogged shortages under inflation. Int J Bus Forecast Mark Intel 3(2):152–164

Wee HM (1997) A replenishment policy for items with a price-dependent demand and a varying rate of deterioration. Prod Planning Control 8(5):494–499

Wu KS (2001) An EOQ inventory model for items with Weibull distribution deterioration, ramp type demand rate, and partial backlogging. Prod Planning Control 12(8):787–793

# Chapter 6
# Inventory Policies for Non-instantaneous Deteriorating Items with Random Start Time of Deterioration

**Nita H. Shah and Pratik H. Shah**

**Abstract** An inventory model for non-instantaneous deteriorating items with random start time of deterioration is investigated in this paper. For many products, the start time of deterioration cannot be predicted due to physical nature of the product. In this paper, products in the inventory system are considered to be deteriorated at a constant rate after a certain random time of inventory received by the retailer. Demand for the product is considered to be price sensitive. Two scenarios viz. with preservation technology investments and without preservation technology investments are compared to obtain retailer's optimal policies which include optimal cycle time, preservation cost, and selling price. The objective is to maximize total profit of retailers with respect to cycle time, selling price, and preservation technology investments. The results indicate that use of preservation technology helps retailers to generate more profit.

**Keywords** Non-instantaneous deterioration · Random start time of deterioration · Preservation technology · Price sensitive demand · Inventory policies

**MSC** 90B05

## 6.1 Introduction

Product demand has been always one of the major concerns for inventory managers. Demand for the product gets affected by various parameters such as stock, time, selling price, quality, different promotional offers, etc. It is very essential to select the precise demand pattern to make optimal inventory decisions. There are certain products for which the demand pattern is very sensitive to the product price. In such demand pattern, notable change can be observed in the product demand as the selling

N. H. Shah (✉) · P. H. Shah
Department of Mathematics, Gujarat University, Ahmedabad, Gujarat 380009, India

P. H. Shah
Department of Mathematics, C.U. Shah Government Polytechnic, Surendranagar, Gujarat 363035, India


N. H. Shah et al. (eds.), *Decision Making in Inventory Management*, Inventory Optimization, https://doi.org/10.1007/978-981-16-1729-4_6

price of the product changes. Increase in selling price is useful in generating revenue but it leads to a decrease in the demand of the product. On the other hand, reduction in product price may attract customers to buy the product but it may be harmful for overall profit of the firm. Deterioration is defined, in general, as spoilage, damage, decay, perishing, fungus, or evaporation of inventory goods. Deterioration of the product may start instant after the production process or it may start later at a certain fixed time or random time after the production. Deterioration affects quality and/or quality of the product that causes loss of goodwill and reduction in the profit of the firm. Inventory managers may use preservation technology to reduce the deterioration rate of product.

Sarkar (2012) investigated an inventory model with delay in payments and time-varying deterioration rate. Dye (2013) studied the effect of preservation technology on a non-instantaneous deteriorating inventory model. Dye and Hsieh (2013) considered instantaneous deterioration with time-dependent demand for inventory model to obtain optimum policies. Hsieh and Dye (2013) gave a production-inventory model incorporating the effect of preservation technology investment where they considered the time fluctuating demand. Shah et al. (2013) gave optimal inventory policies for single-supplier single-buyer deteriorating items with price-sensitive stock-dependent demand and order-linked trade credit. Shah et al. (2021) studied an inventory model for instantaneously deteriorating items with use of preservation technology investments. In development of the model, they considered promotional efforts to promote sales for retailer and quantity discounts from supplier to encourage the retailer for a large order. Singh and Sharma (2013) gave a global optimizing policy for decaying items with ramp type demand considering preservation technology investments. Mishra (2014) studied deteriorating inventory model with controllable deterioration rate for time-dependent demand and time-varying holding cost. Shah and Shah (2014) studied inventory model for deteriorating items with price-sensitive stock-dependent demand under inflation. Tayal et al. (2014) and Zhang et al. (2014) studied inventory model with preservation technology investment with different demand types for instantaneous deteriorating items. Liu et al. (2015) gave joint dynamic pricing and investment strategy for perishable foods with price-quality-dependent demand. Sarkar et al. (2015) investigated inventory model with trade credit policy and variable deterioration for products with maximum lifetime. Singh and Rathore (2015) gave optimum payment policy with preservation technology investment and shortages under trade credit. Tsao (2016) studied joint location inventory and preservation decisions for non-instantaneous deterioration items under delay in payments. Bardhan et al. (2019) considered stock-dependent demand for non-instantaneous deteriorating items.

Most of the researches have been carried out considering instantaneous or non-instantaneous with deterministic start time of deterioration. However, this assumption seems unrealistic. It is not possible to predict exact start time of deterioration. Rahim et al. (2000) considered deterioration starting at a random point to study the inventory model, however they did not consider the idea of preservation technology. Pal et al. (2018) considered deterioration to start at random point with preservation technology where they considered constant demand. Tai et al. (2019) investigated joint inspection and inventory control for deteriorating items with random maximum life time.

In this paper, an inventory model for non-instantaneous deteriorating items with random start time of deterioration is considered. Market demand pattern is considered to be price sensitive. Further, the products are considered to be non-instantaneous deteriorating items with constant rate of deterioration. Retailer may invest in preservation technology to reduce the deterioration rate. With consideration of all these parameters authors aim to maximize total profit and examine optimal decisions for retailer. A numerical example is provided to validate the mathematical model. Moreover, a sensitivity analysis has been carried out to analyze the effect of changes in various inventory parameters on decision variables as well as the total profit, where one inventory parameter is varied by $-20$, $-10$, 10, and 20%. Comparison of both cases 'with preservation' and 'without preservation' have been analyzed to decide which of them is more beneficial for the retailer.

## 6.2 Assumptions and Notations

Authors use the following assumptions and notations in development of mathematical models.

(1) Replenishment rate is infinite and there is no lead time.
(2) The demand rate is $R(p) = \alpha - \beta p$ ; $\alpha, \beta > 0$ where, $\alpha$ is scale demand and $\beta$ is price sensitivity factor.
(3) Products are considered to be non-instantaneous deteriorating with constant rate of deterioration.
(4) Inventory model is for a single cycle $[0, T]$, which includes two phases: (i) $[0, x]$ where there is no deterioration and (ii) $[x, T]$ where products deteriorate at a constant rate. The point in time $x$ at which deterioration starts is a random variable with positive range.
(5) There is no replacement or repair for deteriorated items in the inventory system.
(6) The proportion of reduced deterioration rate $m(\xi)$ is a continuous, concave, increasing function of the retailer's capital investments $\xi$ with $m(0) = 0$ and $\lim_{\xi \to \infty} m(\xi) = 1$. Further, we assume $m'(\xi) > 0$ to ensure that it is worth to invest money in preservation technology and $m''(\xi) < 0$ to ensure diminishing return from capital investments on preservation.

Following notations have been used in the development of the mathematical model.

Decision variables:

$p$ Selling price is \$/unit.
$T$ Cycle length of inventory.
$\xi$ Preservation technology investment in \$/unit.

Other inventory parameters:

| | |
|---|---|
| $A$ | Ordering cost in $/order. |
| $c$ | Cost price in $/unit. |
| $h$ | Holding cost in $ per unit per unit time. |
| $I_1(t)$ | Inventory level during time interval $[0, x]$. |
| $I_2(t)$ | Inventory level during time interval $[x, T]$. |
| $R(p)$ | Price-sensitive demand rate. |
| $x$ | Point in time when deterioration begins, a random variable over $(a, b)$ with $pdf\ f(\cdot)$ and $cdf\ F(\cdot)$. |
| $\theta$ | Deterioration rate $(0 < \theta < 1)$. |
| $m(\xi)$ | Proportion of reduced deterioration rate $(0 \leq m(\xi) \leq 1)$. |

Objective function:

| | |
|---|---|
| $\Pr(T, p, \xi)$ | Average total profit of retailer with preservation technology investments. |

## 6.3 Mathematical Model

Graphical representation of the Inventory model is shown in Fig. 6.1.

Figure 6.1 shows structure of the inventory model. Products in the system are considered as non-instantaneous deteriorating in nature. As per our assumption deterioration starts at random time $x$, hence there is no deterioration in the time interval $[0, x]$ and inventory level decreases due to the demand only, Whereas during $[x, T]$ inventory level is depleted due to combined effect of demand and deterioration.

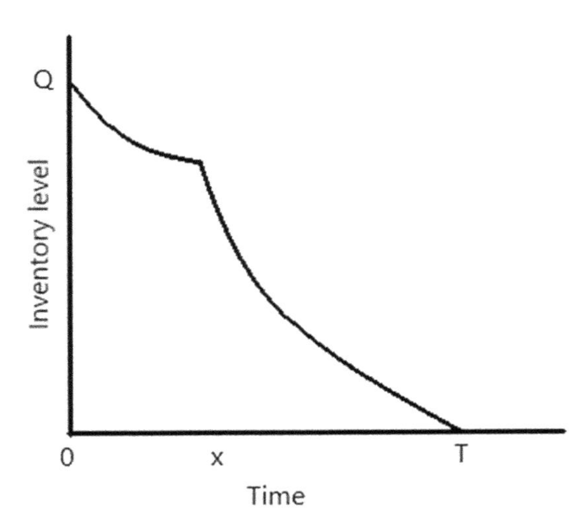


**Fig. 6.1** Graphical representation of the inventory model

Corresponding inventory levels at any point of time t in respective intervals are governed by the differential equations,

$$\frac{d}{dt}I_1(t) = -R(p), \quad 0 \le t \le x \tag{6.1}$$

$$\frac{d}{dt}I_2(t) = -R(p) - (1 - m(\xi)) \cdot \theta \cdot I_2(t), \quad x \le t \le T \tag{6.2}$$

Using the boundary condition $I_2(T) = 0$ and continuity of the demand function for solving Eqs. (6.1) and (6.2) inventory levels $I_1(t)$ and $I_2(t)$ for corresponding time interval can be obtained as below:

For $0 \le t \le x$,

$$I_1(t) = (\alpha - \beta p)\left(x - t + \frac{e^{(1-m(\xi))\theta(t-x)} - 1}{(1 - m(\xi))\theta}\right) \tag{6.3}$$

And for $x \le t \le T$,

$$I_2(t) = (\alpha - \beta p)\left(\frac{e^{(1-m(\xi))\theta(T-t)} - 1}{(1 - m(\xi))\theta}\right) \tag{6.4}$$

The total inventory during the interval [0, $T$] is as given below:

$$I(t) = \int_0^x I_1(t)\mathrm{d}t + \int_x^T I_2(t)\mathrm{d}t \tag{6.5}$$

The ordering quantity is given as,

$$Q = I_1(0) = \int_a^b (\alpha - \beta p)\left(x + \frac{e^{(1-m(\xi))\theta(-x)} - 1}{(1 - m(\xi))\theta}\right) f(x)\mathrm{d}x \tag{6.6}$$

The holding cost is:

$$\mathrm{HC} = h\int_a^b \left[\int_0^x I_1(t)\mathrm{d}t + \int_x^T I_2(t)\mathrm{d}t\right] f(x)\mathrm{d}x \tag{6.7}$$

The total preservation cost is:

$$\mathrm{PTC} = \xi\int_a^b \left[\int_0^x I_1(t)\mathrm{d}t + \int_x^T I_2(t)\mathrm{d}t\right] f(x)\mathrm{d}x \tag{6.88}$$

Total sales revenue is:

$$\mathrm{TR} = p\int\limits_{0}^{T} R(p)t\,\mathrm{d}t \tag{6.9}$$

The total profit of complete inventory cycle $[0, T]$ is given as

$$\Pr(T, p, \xi) = \frac{1}{T}(\mathrm{TR} - A - cQ - \mathrm{HC} - \mathrm{PTC}) \tag{6.10}$$

## 6.4 Numerical Example

Authors now illustrate the inventory model with numerical examples. The objective is to maximize total profit of the retailer which can be obtained by differentiating Eq. (6.10) with respect to decision variables $T$, $p$,$\xi$ and setting them zero in order to get solution. This is shown in the following procedure.

Step 1: Allocate values to all inventory parameters other than decision variables.

Step 2: Work out $\frac{\partial \Pr}{\partial \xi} = 0$, $\frac{\partial \Pr}{\partial p} = 0$ and $\frac{\partial \Pr}{\partial T} = 0$ to get optimum values of decision variables, $T$ $p$ and $\xi$ respectively.

Step 3: Substitute values of decision variables obtained above in Eq. (6.10) to get optimum value of total profit of the retailer.

Consider the following example to validate the mathematical formulation.

***Example*** Let $A = \$\ 5000$ per order, $a = 5$ days, $b = 10$ days, $c = \$\ 30$ per unit, $h = \$\ 5$/unit/day, $\theta = 0.2$, $\alpha = 300$, $\beta = 2$. Demand $R(p) = \alpha - \beta p$ units/day. Authors have considered the reduced deterioration rate $m(\xi) = 1 - \mathrm{e}^{(-k\cdot\xi)}$; where $k = 0.06$ is the simulation coefficient representing the change in the reduced deterioration rate per unit change in capital (Dye 2013). Moreover, authors assume the probability density function of $x$, $f(x) = \begin{cases} \frac{2x}{b^2-a^2} & ;\ a \le x \le b \\ 0 & ;\ otherwise \end{cases}$ where $a = 5 \le x \le 10 = b$, with mean $\mu = 7.7777$ and standard deviation $\sigma = 1.4163$. The form of $pdf$ is selected in such a way that probability of product will start deteriorating, increases with time.

By following the procedure mentioned above to get the optimal values of all the decision variables, optimal values of decision variables and the total profit are obtained as mentioned in Table 6.1 for both scenarios with preservation and without preservation.

Concavity of the profit function can be seen from the following graphs in Fig. 6.2.

Figure 6.2 shows concavity of the profit function with preservation technology investments. Figure 6.2a represents concavity of the profit function with respect to selling price and cycle time, Fig. 6.2b shows that profit function is concave with

**Table 6.1** Optimal values for the inventory model

| Decision variables | Preservation cost ($\xi$) | Retail price ($p$) | Cycle time ($T$) (days) | Total profit |
|---|---|---|---|---|
| With preservation technology investment | \$4.33 | \$85.73 | 28.89 | \$119, 187 |
| Without preservation technology investment | NA | \$88.48 | 20.46 | \$77, 169 |

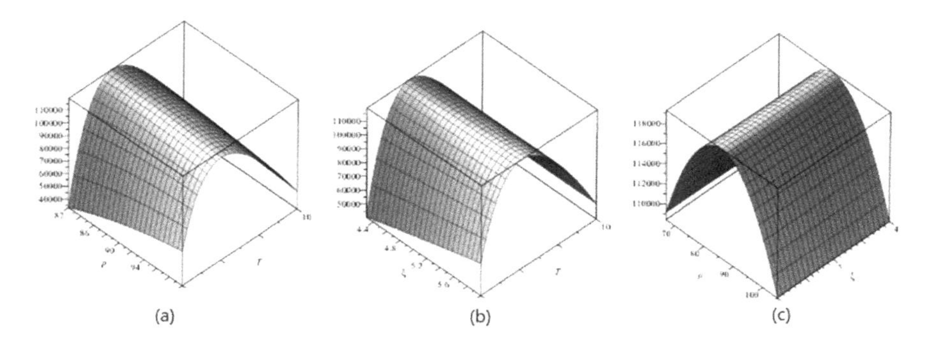


**Fig. 6.2** Concavity of profit function with respect to decision variables (with preservation)

respect to cycle time and preservation cost and Fig. 6.2c depicts concavity of profit function with respect to selling price and preservation cost. Thus, Fig. 6.2 assures the concavity of profit function with respect to all the decision variables.

Next, authors proceed to determine the sensitivity of total profit of retailer, preservation technology cost, cycle time, and selling price with respect to change in other inventory parameters by $-20$, $-10$, 10, and 20% as shown in Table 6.2.

Table 6.2 characterizes sensitivity analysis of decision variables and total profit with respect to change in other inventory parameters. Table 6.2 shows that total profit is very responsive to the parameters $a, b, h, \alpha, \beta, \theta$ and $k$. Parameters $A$ and $c$ have negligible outcome on profit. Increase in $a, b, \alpha$ and $k$ results in increase in total profit. On the other side, the total profit decreases with increase in the parameters $h, \beta$ and $\theta$. Similarly, the sensitivity of preservation cost can be seen in the Table 6.2. Preservation cost increases with increase in the parameters $a$ and $\beta$, while increase in the parameters $b, c, h, \alpha, \theta$ and $k$ reduces preservation technology investments. There is ignorable effect of change in $A$ on preservation cost. Moreover, cycle time is very sensitive to the parameters $a$, $b$, $k, \alpha, \beta$ and $\theta$. Other parameter's effect is negligible to the cycle time. Cycle time increases with increase in $a, b$ and $\alpha$ while it decreases with increase in $h, k, \beta$ and $\theta$. Selling price is very responsive to all the parameters except $A$. It can be observed that with respect to increase in $c$ and $\alpha$,

**Table 6.2** Impact of change in various inventory parameters on decision variables

| Inventory parameters | Decision variables | Percentage change in various inventory parameters | | | | |
|---|---|---|---|---|---|---|
| | | −20% | −10% | 0 | 10% | 20% |
| Ordering cost/order ($A$) | $T$ | 28.8989 | 28.8993 | 28.8998 | 28.9002 | 28.9007 |
| | $P$ | 85.7328 | 85.7333 | 85.7338 | 85.7343 | 85.7348 |
| | $\xi$ | 4.3315 | 4.3303 | 4.3291 | 4.328 | 4.3268 |
| | Profit | 119,222 | 119,205 | 119,187.2 | 119,170 | 119,153 |
| Lower limit of deterioration interval ($a$) | $T$ | 28.4 | 28.65 | 28.8998 | 29.16 | 29.42 |
| | $P$ | 85.94 | 85.84 | 85.7338 | 85.62 | 85.51 |
| | $\xi$ | 4.283 | 4.309 | 4.3291 | 4.343 | 4.351 |
| | Profit | 116,375 | 117,754 | 119,187.2 | 120,666 | 122,181 |
| Upper limit of deterioration interval ($b$) | $T$ | 27.82 | 28.36 | 28.8998 | 29.43 | 29.96 |
| | $P$ | 86.46 | 86.09 | 85.7338 | 85.4 | 85.09 |
| | $\xi$ | 4.72 | 4.53 | 4.3291 | 4.12 | 3.9 |
| | Profit | 112,115 | 115,669 | 119,187.2 | 122,655 | 126,061 |
| Cost price/unit ($c$) | $T$ | 28.9033 | 28.9015 | 28.8998 | 28.8981 | 28.8964 |
| | $P$ | 85.714 | 85.724 | 85.7338 | 85.744 | 85.754 |
| | $\xi$ | 4.355 | 4.342 | 4.3291 | 4.316 | 4.304 |
| | Profit | 119,275 | 119,231 | 119,187.2 | 119,143 | 119,099 |
| Holding cost/unit ($h$) | $T$ | 30.2 | 29.54 | 28.8998 | 28.28 | 27.67 |
| | $P$ | 86 | 85.86 | 85.7338 | 85.6 | 85.47 |
| | $\xi$ | 5.45 | 4.89 | 4.3291 | 3.77 | 3.2 |
| | Profit | 123,528 | 121,329 | 119,187.2 | 117,101 | 115,069 |
| Preservation efficiency scale ($k$) | $T$ | 29.71 | 29.21 | 28.8998 | 28.63 | 28.36 |
| | $P$ | 88.13 | 86.68 | 85.7338 | 85.03 | 84.47 |
| | $\xi$ | 10.82 | 6.75 | 4.3291 | 2.7 | 1.52 |
| | Profit | 113,572 | 116,932 | 119,187.2 | 120,673 | 121,587 |
| Constant demand rate co-efficient ($\alpha$) | $T$ | 28.38 | 28.57 | 28.8998 | 29.26 | 29.61 |
| | $P$ | 70.2 | 77.89 | 85.7338 | 93.65 | 101.6 |
| | $\xi$ | 7.34 | 5.47 | 4.3291 | 3.53 | 2.93 |
| | Profit | 70,199 | 93,026 | 119,187.2 | 148,768 | 181,832 |
| Selling price dependent demand rate co-efficient ($\beta$) | $T$ | 29.79 | 29.3 | 28.8998 | 28.6 | 28.41 |
| | $P$ | 105.59 | 94.54 | 85.7338 | 78.59 | 72.73 |
| | $\xi$ | 2.68 | 3.45 | 4.3291 | 5.35 | 6.59 |
| | Profit | 159,716 | 137,025 | 119,187.2 | 104,809 | 92,968 |
| Natural deterioration rate ($\theta$) | $T$ | 34.14 | 31.23 | 28.8998 | 26.99 | 25.4 |
| | $P$ | 86.72 | 86.2 | 85.7338 | 85.32 | 84.95 |
| | $\xi$ | 4.75 | 4.54 | 4.3291 | 4.13 | 3.94 |
| | Profit | 136,578 | 126,966 | 119,187.2 | 112,744 | 107,306 |

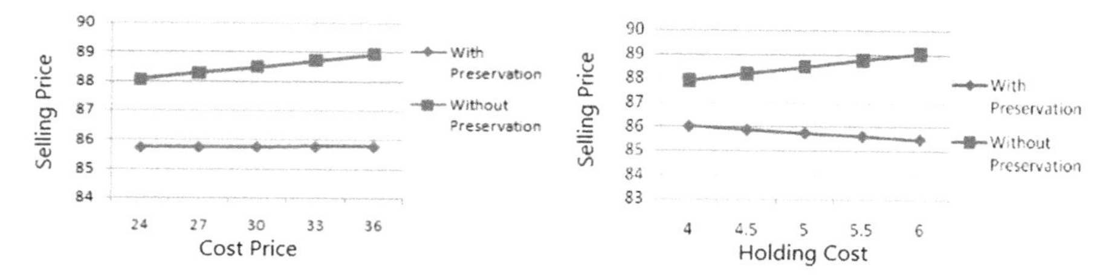


**Fig. 6.3** Sensitivity of selling price with respect to cost price and holding cost (with preservation and without preservation)

selling price also increases. On the other side, increase in $a$, $b$, $h$, $\beta$, $\theta$ and $k$ result in decrease in the selling price. It can be observed from the graph that Total profit and cycle time do not respond significantly to the change in $A$. Profit and cycle time both increases with increase $a$, $b$ and $\alpha$. On the other hand, increase in $c$, $h$, $\beta$ and $\theta$ results in decrease in both the profit and cycle time.

The retailer should wisely decide the investment amount for preservation technology to reduce the deterioration rate so as the total cost does not increase and the total profit can be maximized. Sensitivity analysis of selling price, cycle time, and total profit with preservation technology investments and without preservation technology investments is shown in following Figs. 6.3, 6.4, 6.5, 6.6, 6.7, 6.8 and 6.9.

Figure 6.3 shows that selling price is equally sensitive with respect to cost price in both situations. Increase in cost price gives rise to increase in selling price, which is slightly less in with preservation compared to without preservation case. Similarly, increase in holding cost results the hike in selling price. Selling price without preservation case remains higher than the preservation technology.

Figure 6.4 shows that selling price increases with increase in $\alpha$ and decreases with an increase in $\beta$. This is clearly reflected in the graph above. It can be noted that selling price in with preservation case is slightly less than the selling price in without preservation case.

Figure 6.5 depicts sensitivity of cycle time with respect to cost price and holding cost. First graph represents effect of change in cost price on cycle time and second

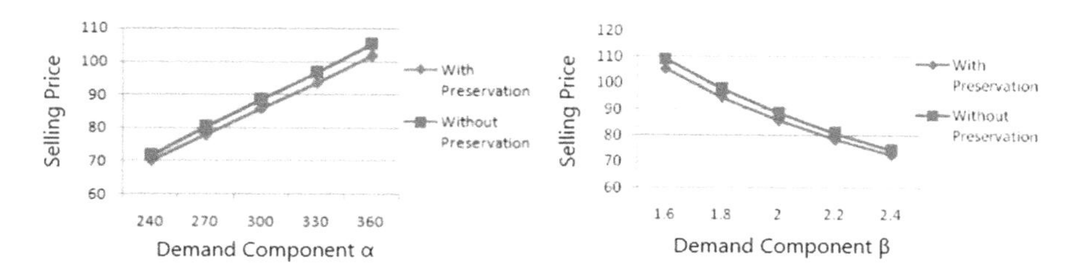


**Fig. 6.4** Sensitivity of selling price with respect to demand components (with preservation and without preservation)

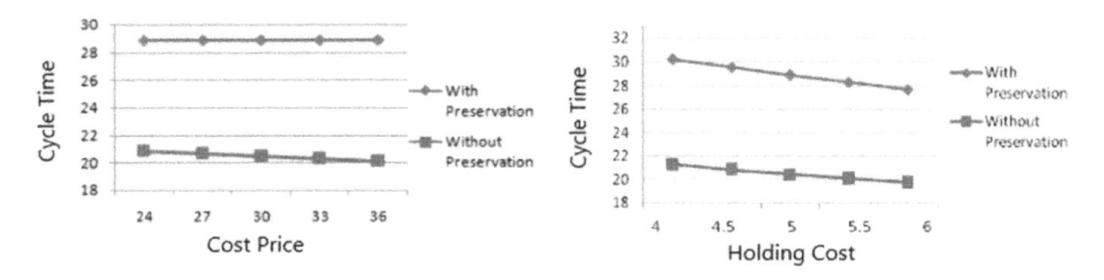


**Fig. 6.5** Sensitivity of cycle time with respect to cost price and holding cost (with preservation and without preservation)

graph shows effect of change in holding cost on cycle time. In both scenarios 'with preservation' and 'without preservation' the cycle time remains almost the same with increase in cost price, while cycle time decreases with increase in holding cost in both cases. Cycle time remains higher in preservation case compared to no-preservation.

Figure 6.6 characterizes the change in cycle time with respect to change in demand components. Cycle time increases with increase in $\alpha$. Cycle time remains higher in preservation case compared to without preservation because the preservation technology let the product last for a longer time. On the other side, increase in $\beta$ results into decrease in the cycle time.

Figure 6.7 represents the effect of change in total profit with respect to cost price and holding cost. First graph represents effect in total profit due to increase in cost price and second graph shows effect of increase of holding cost on total profit.

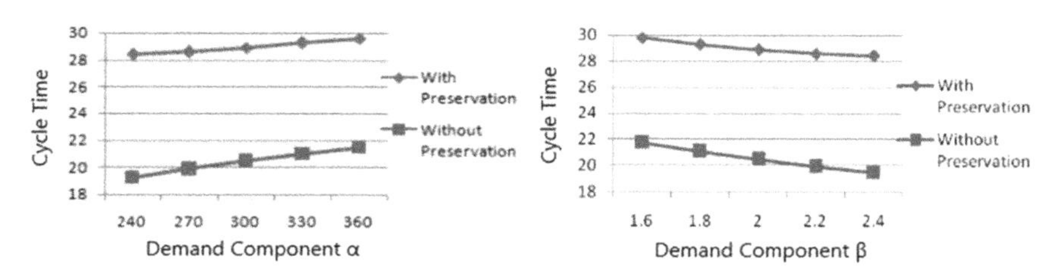


**Fig. 6.6** Sensitivity of cycle time with respect to demand components (with preservation and without preservation)

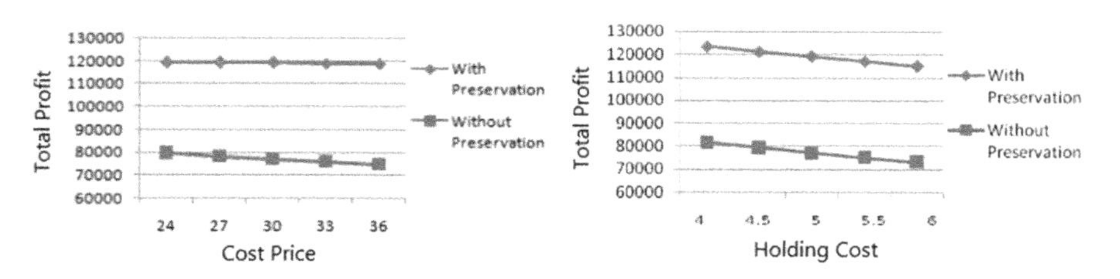


**Fig. 6.7** Sensitivity of total profit with respect to cost price and holding cost (with preservation and without preservation)

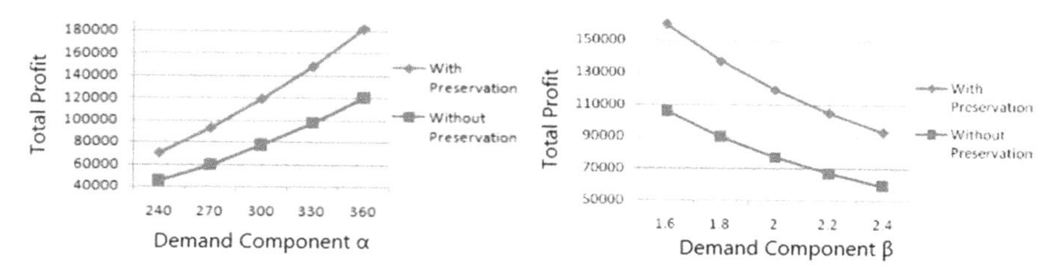


**Fig. 6.8** Sensitivity of Total Profit with respect to demand components (with preservation and without preservation)

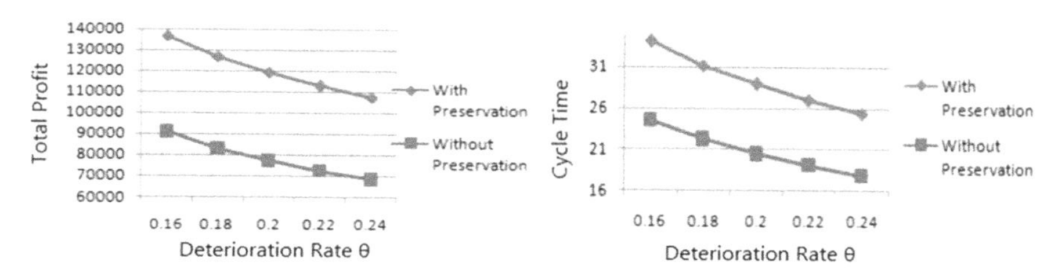


**Fig. 6.9** Sensitivity of Total Profit and Cycle Time with respect to deterioration (with preservation and without preservation)

Increase in the costs associated with inventory results in decrease in total profit which is clearly reflected in both graphs. The total profit in preservation is higher than profit with no preservation.

Figure 6.8 shows how the total profit changes with respect to change in demand components. Total profit increases with the increase in $\alpha$ and decreases with increase in $\beta$ in both the scenario 'with preservation' and 'without preservation'. Total profit is higher in preservation technology case compared to no-preservation case.

Figure 6.9 depicts how the change in the deterioration rate affects the total profit and the cycle time. First graph shows change in total profit with respect to increase in deterioration rate. Here it can be noticed that with increase in the rate of deterioration total profit decreases. However in 'with preservation' due to preservation technology the decrease in total profit is lower compared to 'without preservation' case. Second graph denotes the change in cycle time with respect to increase in deterioration. With increase in the deterioration rate, the cycle time decreases in both scenarios. Due to preservation technology the decrease in 'with preservation' is less than that in 'without preservation'.

## 6.5 Conclusion

Authors have studied non-instantaneous deteriorating products with random start time of deterioration with preservation technology investments. Demand of the product is considered to be price sensitive. Study of the model includes comparison of 'with preservation investments' and 'without preservation investments' through the graphs and detailed analysis has been carried out. The model is validated through numeric example. Sensitivity analysis has been carried out to check the effect of different parameters on decision variables. It is observed from the study that investments in preservation technology give better profit than the non-preservation technology case. However, the retailer needs to take care of investments in preservation technology because it helps in reducing the deterioration rate but the higher amount of investments can increase the capital cost and decrease the total profit.

**Acknowledgements** Authors thank DST-FIST file # MSI-097 for technical support to department of mathematics.

## References

Bardhan S, Pal H, Giri B (2019) Optimal replenishment policy and preservation technology investment for a non-instantaneous deteriorating item with stock-dependent demand. Oper Res Int J 19(2):347–368

Dye CY (2013) The effect of preservation technology investment on a non-instantaneous deteriorating inventory model. Omega 41(5):872–880

Dye CY, Hsieh TP (2013) A practice swarm optimization for solving lot-sizing problem with fluctuating demand and preservation technology cost under trade credit. J Global Optim 55(3):655–679

Hsieh TP, Dye CY (2013) A production-inventory model incorporating the effect of preservation technology investment when demand is fluctuating with time. J Comput Appl Math 239:25–36

Liu G, Zhang J, Tang W (2015) Joint dynamic pricing and investment strategy for perishable foods with price-quality dependent demand. Ann Oper Res 226:397–416

Mishra VK (2014) Deteriorating inventory model with controllable deterioration rate for time dependent demand and time varying holding cost. Yugoslov J Operations Res 24(1):87–98

Pal H, Bardhan S, Giri BC (2018) Optimal replenishment policy for non-instantaneously perishable items with preservation technology and random deterioration start time. Int J Manage Sci Eng Manage 13(3):188–199

Rahim MA, Kabadi SN, Barnerjee PK (2000) A single period perishable inventory model where deterioration begins at a random point in time. Int J Syst Sci 31(1):131–136

Sarkar B (2012) An EOQ model with delay in payments and time varying deterioration rate. Math Comput Model 55(3–4):367–377

Sarkar B, Saren S, Cardenas-Barron LE (2015) An inventory model with trade credit policy and variable deterioration for fixed lifetime products. Ann Oper Res 229:677–702

Shah NH, Shah AD (2014) Optimal cycle time and preservation technology investment for deteriorating items with price-sensitive stock-dependent demand under inflation. J Phys Conf Ser 495(1):12–17

Shah NH, Patel DG, Shah DB (2013) Optimal inventory policies for single-supplier single-buyer deteriorating items with price-sensitive stock-dependent demand and order linked trade credit. Int J Inventory Control Manage 3(1–2):285–301

Shah NH, Shah PH, Patel MB (2021) Retailer's inventory decisions with promotional efforts and preservation technology investments when supplier offers quantity discounts. Opsearch. https://doi.org/10.1007/s-12597-021-00516-6

Singh SR, Rathore H (2015) Optimal payment policy with preservation technology investment and shortages under trade credit. Indian J Sci Technol 8(S7):203–212

Singh SR, Sharma S (2013) A global optimizing policy for decaying items with ramp-type demand rate under two-level trade credit financing taking account of preservation technology. In: Advances in decision sciences, pp 12. Retrieved from https://www.hindawi.com/journals/ads/2013/126385

Tai AH, Xie Y, He W, Ching WK (2019) Joint inspection and inventory control for deteriorating items with random maximum lifetime. Int J Prod Econ 207:144–162

Tayal S, Singh SR, Sharma R, Chauhan A (2014) Two echelon supply chain model for deteriorating items with effective investments in preservation technology. Int J Math Oper Res 6(1):84–105

Tsao YC (2016) Joint location inventory and preservation decisions for non-instantaneous deterioration items under delay in payments. Int J Syst Sci 47(3):572–585

Zhang J, Bai Z, Tang W (2014) Optimal pricing policy for deteriorating items with preservation technology investment. J Ind Manage Optim 10(4):1261–1277

# Chapter 7
# An Inventory Model for Deteriorating Items with Constant Demand Under Two-Level Trade-Credit Policies

**Nita H. Shah, Kavita Rabari, and Ekta Patel**

**Abstract** In today's competitive market, inventory management is a difficult job for every business enterprises. Objects are getting deteriorate after some period of time and result into economic loss. Keeping this in mind, this inventory model is for perishable objects where the rate of deterioration is considered to be constant with a constant demand rate. To reflect the real-life situation, the model explores a two-level trade-credit policy, i.e. the supplier offers certain credit period to the retailer and simultaneously the retailer permits a permissible delay in payment to the consumers that helps to increase the demand. If the retailer clears its entire amount during the end of first credit period, then the retailer can utilize it to earn interest. Moreover, if the retailer fails to clear the account by the end of first period, then he/she is allowed to pay off the balance after first credit period or by the end of second credit period. Here, the financial loans can be reduced through constant demand and interest earned. This paper uses a classical optimization method and calculated several numerical examples to elaborate the model. Convexity of cost function is proved through graphs. The objective of the paper is to minimize the total cost with respect to the inventory cycle time. At last, sensitivity analysis is done to study the effects of varying inventory parameters on decision variable and optimal solution.

**Keywords** Constant deterioration · Constant demand rate · Two level trade-credit · Cycle time · Sensitivity

**MSC** 90B05

## 7.1 Introduction

With the rapid development of competition and technology between the business enterprises, companies are feeling the necessity of inventory models as a decision-making device for developing their business effectively. It is well-known fact that a

---

N. H. Shah (✉) · K. Rabari · E. Patel
Department of Mathematics, Gujarat University, Ahmedabad, Gujarat 380009, India



N. H. Shah et al. (eds.), *Decision Making in Inventory Management*,
Inventory Optimization, https://doi.org/10.1007/978-981-16-1729-4_7

general model always represents an enhanced outcome in provisions of maximizing the profit or minimizing the total cost. The traditional inventory models are based on the fact that a retailer has to pay as soon as he received the product. However, this may not be correct. In real-life situations, it is common to observe that the supplier will offer certain time period to the retailer to settle the account. The term is known as trade-credit policy. The company often uses this policy to promote the products. Generally, if the amount is paid before the permissible period, the interest charge is zero and the retailer can use the sale revenue to earn interest. Nevertheless, if the retailer is not able to pay off the amount within the permissible period, an interest is charged by the supplier. This brought up economic advantage to the retailers as they can make some interest from the proceeds. Hence, this model develops a two-level credit policy to reflect a real-life situation. Also, items like fruits, vegetables, dairy products, etc. are perishable with time. It results in loss of marginal values of products, loss of profit and loss of goodwill that lead to reduce the usefulness of the product. Therefore, deterioration plays a vital role and cannot be ignored. Together with rate of deterioration, demand is also one of the factors that influence the sale a lot. Here, in this model instead of taking demand dependent on some specific parameters, to establish model for general cases, the demand rate is considered to be constant. By keeping this in mind, an inventory model is developed with constant demand and deterioration rate. This paper calculates the inventory cycle time where total cost is minimized with respect to decision variables. The paper is structured as follows: Sect. 7.2 is literature review. In Sect. 7.3, notations and assumptions are introduced that are used in proposed model. The inventory model is formulated in Sect. 7.4. Section 7.5 contains computational algorithm. Section 7.6 describes several numerical examples together with sensitivity analysis with respect to inventory parameters. At last, Sect. 7.7 provides conclusion with future scope.

## 7.2 Literature Review

The trade-credit policy is widely used in inventory models to increase the sale of commodities and to attract more customers. Teng and Chang (2009) developed an EPQ model for two-level credit policy and develop some appropriate results for obtaining optimal solution. Wu et al. (2014) proposed a model for deteriorating items having date of expiration with two-level credit policy. They proved not only the existence of optimal cycle time and trade credit but also the uniqueness of the solution under some numerical examples to modify the problem. Cheng and Kang (2010) developed an integrated model with delay in payment. The model considers vendor–buyer and buyer–customer relationship and presented a price-negotiation scheme to allocate the increased amount of profit. Chung et al. (2014) established an economic production quantity model for exponentially perishable objects under two-level credit periods. The objective is to determine optimal replenishment policy to minimize the relevant cost. Teng et al. (2013) provided a linear non-decreasing demand function of time under permissible credit period. Sarkar et al. (2015) introduced a model for variable deterioration with fixed lifetime products. In this model, numerical examples are illustrated with graphical representation. Sarkar et al. (2013) developed a model for perishable objects where the demand is considered to be time dependent. The objective is to maximize the total profit. Shah et al. (2013) proposed an inventory model with non-instantaneous deteriorating item. In this model, the demand rate is assumed to be a function of selling price and advertisement of an item. Chung (2011) developed an inventory model for two-level credit period policy. The condition that interest charged should be greater than the interest earned is relaxed in this model. Tsao et al. (2011) focused on the production problem under credit policy and reworking of imperfect items. Sang and Tripathi (2012) proposed an EOQ model for deteriorating items with constant demand. The production rate is demand sensitive. Due to continuous demand, shortages occurred and are completely backlogged. Srivastava and Gupta (2007) developed an infinite time-horizon inventory model for perishable objects assuming constant and time-dependent demand rate. Khanra et al. (2011) developed an EOQ model for deteriorating item with trade-credit policy where the demand is time dependent. Sarkar and Sarkar (2013) introduced a model for deteriorating item under stock-dependent demand. They consider backlogging rate and deterioration as time-varying function. Mishra et al. (2013) gave an inventory model for deteriorating items having time-dependent demand and holding cost. The model permits partial backlogging due to shortages.

Tripathi and Mishra (2012) developed an inventory model for constant demand and constant deterioration rate under trade credit. Skouri et al. (2011) proposed a model for ramp-type demand rate where the deterioration rate is constant and the unsatisfied demand is partially backlogged. The model allows permissible delays in payment to attract the customers. Shah (2017) developed a three-layered integrated inventory model for perishable objects under two-level trade-credit policy having quadratic demand. Lio et al. (2018) developed a two-level trade-credit policy for finding feasible order quantity. Cardenas et al. (2020) proposed an EOQ model for nonlinear stock-dependent holding cost in which stock-dependent demand is to be considered. Tiwari et al. (2020) analysed an optimal ordering policy for deteriorating items by assuming complete backlogging. Yang (2004) developed an EOQ model using quantity discount. Li (2014) suggested an optimal control production model under permissible tradable emission. Yang (2019) studied an inventory model for deteriorating items under two-level trade credit with limited storage capacity. This model extends the existing literature of inventory models for deteriorating objects. As with credit periods, deterior̃ation is the key factor that influences the objective function directly.

## 7.3 Notations and Assumptions

The model is formulated using following notations and assumptions.

**Notations**

| | |
|---|---|
| $R$ | Constant demand rate |
| $\theta$ | Constant rate of deterioration |
| $h$ | Inventory holding cost per unit (dollars/unit) |
| $C$ | Purchase cost per unit (dollars/unit) |
| $p$ | Selling price per unit (dollars/unit) |
| $M$ | The first credit period by the supplier to the retailer's without spare charges |
| $N$ | Second credit period with an interest of $I_2$, where $N > M$ |
| $A$ | Ordering cost per order(dollars/order) |
| $I_{c1}$ | Interest charge per unit per year during time interval $[M, N]$ (dollars/year) |
| $I_{c2}$ | Interest charge per unit per year during time interval $[N, T]$ (dollars/year) |
| $I_e$ | Interest earned per unit per year (dollars/year) |
| $T$ | Cycle time (in years) |
| $I(t)$ | Inventory level during time $[0, T]$ |
| $\mathrm{TC}(T^*)$ | Total cost per year (dollars/year) |
| $W_1$ | $\frac{p}{C}M + \frac{pI_e}{2C}N^2$ |
| $W_2$ | $\frac{p}{C}N + \frac{pI_e}{2C}\left(M^2 + (N-M)^2\right)$ |

**Assumptions**

- Demand rate for object is constant with time.
- Shortages are not permissible.
- Replenishment rate is infinite.
- For $M > T$, the rate of interest charge is zero and the retailers earn some interest on sales revenue by the time $M$.
- For $M \leq T$, different cases are possible. Initially, if the retailer clears the account by $M$, the interest charge is zero and he/she can earn interest of $I_e$ on sales revenue throughout the cycle time $T$. Secondly, if the retailer is unable to pay up to $M$ or before time period $N$, then the supplier charges an interest of $I_{c1}$ on the retailer and also utilizes the sales revenue to clear the unpaid amount. Lastly, if the retailer pays after time period $N$, an interest of $I_{c2}$ is charged on retailer.

## 7.4 Mathematical Model

In this section, the model is formulated for two-level trade-credit policies for constant demand and deterioration rate. Initially at $t = 0$, the production rate is $Q$ given by $\frac{R}{\theta}\left(e^{\theta T} - 1\right)$, where the demand is considered to be constant, i.e. $R$. During time interval $[0, T]$, the inventory level decreases due to the effect of deterioration and customer's consumption rate and reaches zero at the end of cycle time $t = T$.

The differential equation of the inventory system for the time $[0, T]$ is given by

$$\frac{\mathrm{d}I(t)}{\mathrm{d}t} = -\theta I(t) - R \tag{7.1}$$

Using boundary condition $I(T) = 0$, the inventory level is

$$I(t) = \frac{R}{\theta}\left(e^{\theta(T-t)} - 1\right) \tag{7.2}$$

The order quantity $Q$ is obtained using initial condition $I(0) = Q$ and is given by

$$Q = \frac{R}{\theta}\left(e^{\theta T} - 1\right) \tag{7.3}$$

The costs comprising of the total annual cost are listed below:

- Ordering cost (OC) $= \frac{A}{T}$
- Holding cost (HC) $= \frac{h}{T}\int\limits_{0}^{T} I(t)\mathrm{d}t$
- Related to the last two assumption, there are four different cases with respect to interest earned and interest charged per year

Case 1: $T \leq M$.

Following Liao et al. (2018), in this case, the credit period $M$ is greater than the cycle time $T$, and the retailers sold out all the products before the permissible credit period. So, the interest charge on retailer is zero.

Interest charge $(IC_1) = 0$.

There are two different elements for the interest earned as mentioned bellow:

Firstly, the interest earned by the retailer during time interval $[0, M]$ is

Interest earned (IE11) $= \frac{pI_e}{T}\int\limits_{0}^{T} Rt\mathrm{d}t$.

Secondly, the interest earned during time $[M, T]$ is

Interest earned (IE12) $= \frac{1}{T}\left(pI_eRT + \frac{pRT^2I_e^2}{2}\right)(M - T)$.

Therefore, the total interest earned is given by

Interest earned

$(\mathrm{IE}_1) = \mathrm{IE}11 + \mathrm{IE}12 = \frac{pI_e}{T}\int\limits_{0}^{T} Rt\mathrm{d}t + \frac{1}{T}\left(pI_eRT + \frac{pRT^2I_e^2}{2}\right)(M - T)$.

Case 2: $M < T \leq W_1$.

Here, as $T \leq W_1$ it means the retailers clear all its account up to $M$. Hence, the interest charge is zero.

Interest charge $(\mathrm{IC}_2) = 0$.

There are three different elements for interest earned as follows:

Firstly, interest is earned by the retailer on sales revenue during time $[0, M]$.

Interest earned (IE21) $= \frac{pI_e}{T} \int_0^M Rt\,\mathrm{d}t$.

Secondly, the retailer earns interest on sales revenue due to the sale up to cycle time $T$.

Interest earned (IE22) $= \frac{pI_e}{T} \int_0^{T-M} Rt\,\mathrm{d}t$.

Lastly, the interest earned by the retailer on the sales revenue during time $[M, T]$.

Interest earned (IE23) $= \frac{I_e}{T}\left(p\int_0^M R\mathrm{d}t + pI_e\int_0^M Rt\mathrm{d}t - \mathrm{CRT}\right)(T-M)$.

So, the total interest earned in this case is given by

Interest earned

$$(\mathrm{IE}_2) = \mathrm{IE21} + \mathrm{IE22} + \mathrm{IE23} = \left(\begin{array}{l} \frac{pI_e}{T}\int_0^M Rt\mathrm{d}t + \frac{pI_e}{T}\int_0^{T-M} Rt\mathrm{d}t \\ +\frac{I_e}{T}\left(p\int_0^M R\mathrm{d}t + pI_e\int_0^M Rt\mathrm{d}t - \mathrm{CRT}\right)(T-M) \end{array}\right)$$

Case 3: $W_1 < T \le W_2$.

Here, as $W_1 < T$ that is, the sales revenue achieved by the retailer is less than the purchase cost up to time period $M$. Also, $T \le W_2$ which means the retailer decides to decrease the loan amount by the demand and sales revenue and decides to clear its entire purchase amount up to $N$ or before that. The unpaid balance is given by

Unpaid balance $(U_1) = \mathrm{CQ} - p\int_0^M R\mathrm{d}t - pI_e\int_0^M Rt\mathrm{d}t$.

Charges applied on unpaid balance with an interest rate of $I_{C1}$ for time $M$.

Interest charge $\mathrm{IC}_3 = \frac{I_{C1}U_1^2}{pQ}\int_0^{T-M} I(t)\mathrm{d}t$.

Interest earned for this case is as follows:

Firstly, Interest earned by retailers on sales revenue for time $[0, M]$ is.

Interest earned (IE31) $= \frac{pI_e}{T}\int_0^M Rt\mathrm{d}t$.

Secondly, the retailer uses the sales revenue to gross interest throughout the time from

$M + \frac{U_1}{p\alpha}$ to $T$. So, interest earned is given by

Interest earned (IE32) $= \frac{pI_e}{T}\int_{M+\frac{U_1}{Rp}}^{T} Rt\mathrm{d}t$.

Hence, the total interest earned is given by

Interest earned $(\mathrm{IE}_3) = \mathrm{IE31} + \mathrm{IE32} = \frac{pI_e}{T}\int_0^M Rt\mathrm{d}t + \frac{pI_e}{T}\int_{M+\frac{U_1}{Rp}}^{T} Rt\mathrm{d}t$.

Case 4: $W_2 < T$.

Here in this case, the retailer is unable to clear the account at $M$ and decided to pay it after $N$.

For interest charges, there are two different elements as follows:

Firstly, the supplier charges an interest of $I_1$ on unpaid balance $U_1$ during time $[M, N]$ is

Interest charged (IC41) $= \frac{I_{c1}(N-M)U_1}{T}$.

Secondly, charge on the unpaid balance with rate of $I_{c2}$ at time period $N$. Unpaid balance is

Unpaid balance $(U_2) = \text{CQ} - pI_e \int_0^M Rt\text{d}t - pI_e \int_0^{N-M} Rt\text{d}t - p\int_0^M R\text{d}t - p\int_0^{N-M} R\text{d}t$, the interest charge

Interest charge (IC42) $= \frac{I_{c2}U_2^2}{pQ}\int_N^T I(t)\text{d}t$.

The total interest charge is

Interest charge $(\text{IC}_4) = \text{IC41} + \text{IC42} = \frac{I_{c1}(N-M)U_1}{T} + \frac{I_{c2}U_2^2}{pQ}\int_N^T I(t)\text{d}t$.

The interest earned is given by

Interest earned $(\text{IE}_4) = \frac{pI_e}{T}\int_0^M Rt\text{d}t$.

The total annual cost related to the different cases is mentioned below:

$$\text{TC}_1(T) = (\text{OC} + \text{HC} + \text{IC}_1 - \text{IE}_1),\ \text{for } T \le M \tag{7.4}$$

$$\text{TC}_2(T) = (\text{OC} + \text{HC} + \text{IC}_2 - \text{IE}_2),\ \text{for } M < T \le W_1 \tag{7.5}$$

$$\text{TC}_3(T) = (\text{OC} + \text{HC} + \text{IC}_3 - \text{IE}_3),\ \text{for } W_1 < T \le W_2 \tag{7.6}$$

$$\text{TC}_4(T) = (\text{OC} + \text{HC} + \text{IC}_4 - \text{IE}_4),\ \text{for } W_2 < T \tag{7.7}$$

where total cost is given below. Here $T_i^*$ denotes the optimal cycle time for $\text{TC}_i(T)$ on $T > 0$ if $T_i^*$ exists for $i = 1$ to 4.

$$\text{TC}(T*) = \max\{\text{TC}_1(T_1^*), \text{TC}_2(T_2^*), \text{TC}_3(T_3^*), \text{TC}_4(T_4^*)\} \tag{7.8}$$

Here,

$$\text{TC}(T_1^*) = \max\{\text{TC}_1(T) : 0 < T \le M\} \tag{7.9}$$

$$\text{TC}(T_2^*) = \max\{\text{TC}_2(T) : M < T \le W_1\} \tag{7.10}$$

$$\mathrm{TC}(T_3^*) = \max\{\mathrm{TC}_3(T) : W_1 < T \leq W_2\} \tag{7.11}$$

and

$$\mathrm{TC}(T_4^*) = \max\{\mathrm{TC}_4(T) : W_2 \leq T\} \tag{7.12}$$

## 7.5 Computational Algorithm

The model uses classical optimization method. The goal is to minimize the total cost for the inventory model. The algorithm is based on the preceding steps.

Step 1: Assign numerical values to the inventory parameters.

Step 2: Compute first-order partial derivative for $\mathrm{TC}_1(T), \mathrm{TC}_2(T), \mathrm{TC}_3(T), \mathrm{TC}_4(T)$ with respect to the decision variable $T$ and equating them to zero.

$$\frac{\partial \mathrm{TC}_1(T)}{\partial T} = 0, \frac{\partial \mathrm{TC}_2(T)}{\partial T} = 0, \frac{\partial \mathrm{TC}_3(T)}{\partial T} = 0, \frac{\partial \mathrm{TC}_4(T)}{\partial T} = 0 \tag{7.13}$$

An optimal value of decision variable is $T$ obtained using Eq. (7.13). Hence, the total cost for all the cases can be solved using Eqs. (7.4) to (7.7), and the optimal total annual cost is the one that is satisfied by Eq. (7.8) that also satisfies the respective condition.

Step 3: Convexity of total annual cost is confirmed by means of graphs.

## 7.6 Numerical Example and Sensitivity Analysis

***Example 1*** For $\alpha = 9000$, $C = \$4$/unit, $p = \$20$/unit, $\theta = 0.1$, $h = \$2$/unit/year, $A = \$200$/order, $I_e = \$0.11/\$$/year, $I_{c1} = \$0.14/\$$/year, $I_{c2} = \$0.20/\$$/year, $M = 0.15$ year, $N = 0.2$ year. Using the above procedure, the optimal decision variable is $T_1^* = 0.103$ year that gives $\mathrm{TC}(T^*) = \$916.31$.

***Example 2*** For $\alpha = 400$, $C = \$40$/unit, $p = \$50$/unit, $\theta = 0.2$, $h = \$38$/unit/year, $A = \$156$/order, $I_e = \$0.04/\$$/year, $I_{c1} = \$0.05/\$$/year, $I_{c2} = \$0.06/\$$/year, $M = 0.12$ year, $N = 0.15$ year. Using the above procedure, the optimal decision variable is $T_2^* = 0.140$ year that gives $\mathrm{TC}(T^*) = \$2145.23$.

***Example 3*** For $\alpha = 120$, $C = \$13$/unit, $p = \$14$/unit, $\theta = 0.2$, $h = \$4$/unit/year, $A = \$8$/order, $I_e = \$0.04/\$$/year, $I_{c1} = \$0.05/\$$/year, $I_{c2} = \$0.09/\$$/year, $M = 0.15$ year, $N = 0.17$ year. Using the above procedure, the optimal decision variable is $T_3^* = 0.170$ year that gives $\mathrm{TC}(T^*) = \$83.88$.

***Example 4*** For $\alpha = 80$, $C = \$13$/unit, $p = \$13.001$/unit, $\theta = 0.02$, $h = \$4$/unit/year, $A = \$0.8$/order, $I_e = \$0.005$/\$/year, $I_{c1} = \$0.0051$/\$/year, $I_{c2} = \$0.0052$/\$/year, $M = 0.05$ year, $N = 0.05001$ year. Using the above procedure, the optimal decision variable is $T_4^* = 0.070$ year that gives $\mathrm{TC}(T^*) = \$22.55$.

Convexity of optimal solutions through graphs is shown below:

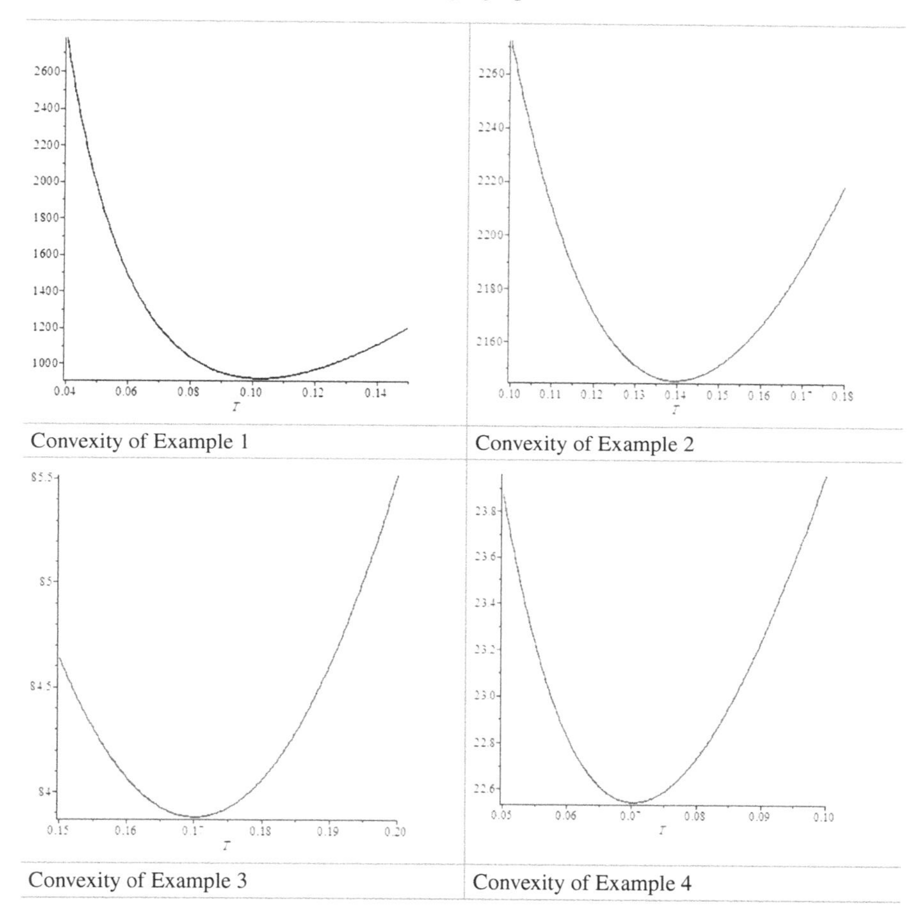


Convexity of Example 1

Convexity of Example 2

Convexity of Example 3

Convexity of Example 4

Also, some numerical results depending on the different cases are discussed below in Table 7.1.

A sensitivity analysis is done that represents the impact of changing inventory parameters by −20%, −10%, +10% and +20% on decision variable and on the total cost. Here, an analysis is performed for the fourth case as it includes all the cases.

Following results are being observed through Table 7.2.

- The more will be the demand rate, less will be the optimal cycle time and larger will be the total cost. Here, the increase is not beneficial as it increases the total cost.

**Table 7.1** Optimal solutions for different cases with data

| Parameters and decision variable | 1 | 2 | 3 | 4 | 5 | 6 |
|---|---|---|---|---|---|---|
| $\alpha$ | 9000 | 120 | 104 | 400 | 190 | 198 |
| $C$ | 4 | 13 | 25 | 40 | 45 | 44 |
| $p$ | 20 | 14 | 26 | 50 | 45.001 | 44.0001 |
| $\theta$ | 0.1 | 0.2 | 0.025 | 0.2 | 0.01 | 0.01 |
| $h$ | 2 | 4 | 6 | 38 | 7 | 7 |
| $A$ | 200 | 8 | 10 | 156 | 10 | 10 |
| $I_e$ | 0.11 | 0.04 | 0.01 | 0.04 | 0.0075 | 0.00005 |
| $I_{C1}$ | 0.14 | 0.05 | 0.012 | 0.05 | 0.01001 | 0.0100005 |
| $I_{C2}$ | 0.20 | 0.09 | 0.015 | 0.06 | 0.010011 | 0.0100051 |
| $M$ | 0.15 | 0.15 | 0.10 | 0.12 | 0.114 | 0.12001 |
| $N$ | 0.2 | 0.17 | 0.12 | 0.15 | 0.11401 | 0.12009 |
| $W_1$ | 0.761 | 0.162 | 0.104 | 0.151 | 0.11405 | 0.12001 |
| $W_2$ | 1.007 | 0.184 | 0.125 | 0.188 | 0.11406 | 0.12009 |
| $T$ | $T_1^* = 0.103$ | $T_3^* = 0.17$ | $T_4^* = 0.10$ | $T_2^* = 0.14$ | $W_2^* =$ 0.11406 | $W_1^* =$ 0.12009 |
| Optimal solutions | | | | | | |
| TC($T*$) | 916.31 | 83.88 | 11.38 | 2145.23 | 159.7 | 166.5 |

- With an increase in purchase cost, the demand for the products due to high rate decreases that influences the cycle time. The cycles' time increases, and it is obvious that the total cost increases. Therefore, the increase is not advisable.
- Increase in selling price reduces the total cost. So, the change is acceptable.
- Higher deterioration rate forces the retailer to invest more. The change is not preferable as it increases the optimality cost.
- The impact is negative as holding cost and ordering cost are the key factors that directly influence the budgets of a company. With an increase in these parameters, the total cost increases.
- An increase in the total cost decreases with a decrease in cycle time.
- With an increase in parameters, total cost increases.
- Credit periods help to boost the products demand. Here, the increase is not sensible as it increases the total cost.

## 7.7 Conclusion

The paper develops an inventory model for constant demand and constant rate of deterioration. The model considers objects that are getting expired, spoilt and deteriorated with respect to time. It plays a crucial role in environment of marketplace, and

**Table 7.2** Sensitivity analysis

| Parameters | Values of parameters | $T$ | Optimal solutions $\text{TC}(T*)$ |
|---|---|---|---|
| $\alpha$ | 32 | 0.264 | 59.12 |
| | 36 | 0.253 | 61.98 |
| | 40 | 0.243 | 64.64 |
| | 44 | 0.235 | 67.13 |
| | 48 | 0.228 | 69.48 |
| $C$ | 23 | 0.243 | 64.64 |
| | 25.3 | 0.234 | 66.31 |
| | 27.6 | 0.226 | 67.96 |
| $p$ | 23.4 | 0.237 | 66.00 |
| | 26 | 0.243 | 64.64 |
| $\theta$ | 0.016 | 0.2432 | 64.62 |
| | 0.018 | 0.2431 | 64.63 |
| | 0.02 | 0.2430 | 64.64 |
| $h$ | 3.6 | 0.246 | 62.68 |
| | 4 | 0.243 | 64.64 |
| | 4.4 | 0.240 | 66.57 |
| | 4.8 | 0.237 | 68.49 |
| $A$ | 8 | 0.229 | 56.17 |
| | 9 | 0.236 | 60.47 |
| | 10 | 0.243 | 64.64 |
| $I_e$ | 0.18 | 0.2431 | 64.64 |
| | 0.198 | 0.2427 | 64.39 |
| | 0.216 | 0.242 | 64.13 |
| $I_{c1}$ | 0.167 | 0.244 | 64.21 |
| | 0.185 | 0.243 | 64.64 |
| $I_{c2}$ | 0.152 | 0.249 | 64.16 |
| | 0.171 | 0.246 | 64.41 |
| | 0.19 | 0.243 | 64.64 |
| $M$ | 0.072 | 0.2428 | 63.05 |
| | 0.08 | 0.2426 | 61.52 |
| $N$ | 0.108 | 0.236 | 64.46 |
| | 0.12 | 0.243 | 64.64 |

the loss that occurs due to that cannot be ignored. The marginal insight shows that higher deterioration rate forces the retailer to spend more. So, it is always preferable to invest more in reducing the rate of deterioration. Also, permissible credit period helps to boost the demand. This model considers a two-level credit period. The model evaluates the optimal cycle time under different cases while minimizing the total cost. The sensitivity analysis shows that the total relevant cost is sensitive to selling price and credit periods. A small change in these parameters highly fluctuates the total cost. The model uses classical optimization method for solution procedure. The present work can be expanded in different ways. One can consider demand to be time dependent or stock dependent with variable deterioration rate having variable holding cost. For smooth business, the companies may offer discounts to attract the customers. To control the rate of deterioration, preservation technology investments are made.

**Acknowledgements** Second author (Kavita Rabari) is funded by a Junior Research Fellowship from the Council of Scientific & Industrial Research (file no.-09/070(0067)/2019-EMR-I), and all the authors are thankful to DST-FIST file # MSI-097 for technical support to the Department of Mathematics, Gujarat University.

## References

Cárdenas-Barrón LE, Shaikh AA, Tiwari S, Treviño-Garza G (2020) An EOQ inventory model with nonlinear stock dependent holding cost, nonlinear stock dependent demand and trade credit. Comput Industr Eng 139:105557

Chen LH, Kang FS (2010) Integrated inventory models considering the two-level trade credit policy and a price-negotiation scheme. Eur J Oper Res 205(1):47–58

Chung KJ (2011) The simplified solution procedures for the optimal replenishment decisions under two levels of trade credit policy depending on the order quantity in a supply chain system. Expert Syst Appl 38(10):13482–13486

Chung KJ, Cárdenas-Barrón LE, Ting PS (2014) An inventory model with non-instantaneous receipt and exponentially deteriorating items for an integrated three layer supply chain system under two levels of trade credit. Int J Prod Econ 155:310–317

Khanra S, Ghosh SK, Chaudhuri KS (2011) An EOQ model for a deteriorating item with time dependent quadratic demand under permissible delay in payment. Appl Math Comput 218(1):1–9

Li S (2014) Optimal control of the production–inventory system with deteriorating items and tradable emission permits. Int J Syst Sci 45(11):2390–2401

Liao JJ, Huang KN, Chung KJ, Lin SD, Ting PS, Srivastava HM (2018) Mathematical analytic techniques for determining the optimal ordering strategy for the retailer under the permitted trade-credit policy of two levels in a supply chain system. Filomat 32(12):4195–4207

Mishra VK, Singh LS, Kumar R (2013) An inventory model for deteriorating items with time-dependent demand and time-varying holding cost under partial backlogging. J Industr Eng Int 9(1):4

Roy SK, Pervin M, Weber GW (2018) A two-warehouse probabilistic model with price discount on backorders under two levels of trade-credit policy. J Industr Manag Optim 13(5):1

Sarkar B, Sarkar S (2013) An improved inventory model with partial backlogging, time varying deterioration and stock-dependent demand. Econ Model 30:924–932

Sarkar B, Saren S, Wee HM (2013) An inventory model with variable demand, component cost and selling price for deteriorating items. Econ Model 30:306–310

Sarkar B, Saren S, Cárdenas-Barrón LE (2015) An inventory model with trade-credit policy and variable deterioration for fixed lifetime products. Ann Oper Res 229(1):677–702

Shah NH (2017) Three-layered integrated inventory model for deteriorating items with quadratic demand and two-level trade credit financing. Int J Syst Sci: Oper Logist 4(2):85–91

Shah NH, Soni HN, Patel KA (2013) Optimizing inventory and marketing policy for non-instantaneous deteriorating items with generalized type deterioration and holding cost rates. Omega 41(2):421–430

Skouri K, Konstantaras I, Papachristos S, Teng JT (2011) Supply chain models for deteriorating products with ramp type demand rate under permissible delay in payments. Expert Syst Appl 38(12):14861–14869

Srivastava M, Gupta R (2007) EOQ Model for deteriorating items having constant and time-dependent demand rate. Opsearch 44(3):251–260

Teng JT, Chang CT (2009) Optimal manufacturer's replenishment policies in the EPQ model under two levels of trade credit policy. Eur J Oper Res 195(2):358–363

Teng JT, Yang HL, Chern MS (2013) An inventory model for increasing demand under two levels of trade credit linked to order quantity. Appl Math Model 37(14–15):7624–7632

Tiwari S, Cárdenas-Barrón LE, Shaikh AA, Goh M (2020) Retailer's optimal ordering policy for deteriorating items under order-size dependent trade credit and complete backlogging. Comput Industr Eng 139:105559

Tripathi RP, Misra SS (2012) An optimal inventory policy for items having constant demand and constant deterioration rate with trade credit. Int J Inform Syst Supply Chain Manage (IJISSCM) 5(2):89–95

Tripathi RP, Sang N (2012) EOQ model for constant demand rate with completely backlogged and shortages. J Appl Comput Math 1(121):2

Tsao YC, Chen TH, Huang SM (2011) A production policy considering reworking of imperfect items and trade credit. Flex Serv Manuf J 23(1):48–63

Wu J, Ouyang LY, Cárdenas-Barrón LE, Goyal SK (2014) Optimal credit period and lot size for deteriorating items with expiration dates under two-level trade credit financing. Euro J Oper Res 237(3):898–908

Yang PC (2004) Pricing strategy for deteriorating items using quantity discount when demand is price sensitive. Eur J Oper Res 157(2):389–397

Yang HL (2019) Optimal ordering policy for deteriorating items with limited storage capacity under two-level trade credit linked to order quantity by a discounted cash-flow analysis. Open J Bus Manage 7(02):919

# Chapter 8
# Supply Chain Coordination for Deteriorating Product with Price and Stock-Dependent Demand Rate Under the Supplier's Quantity Discount

**Chetan A. Jhaveri and Anuja A. Gupta**

**Abstract** In this research paper, optimal ordering and pricing strategy for deteriorating products is developed when demand of a product depends on selling price and stock availability. Without supply chain coordination, the buyer makes policy to maximize its own profit which may not be beneficial to the vendor. Vendor can offer quantity discount as an incentive to encourage buyer to participate in the coordinated strategy. To coordinate the vendor–buyer decisions, two coordination policies are presented in this paper. First, coordinated supply chain strategy is developed to show that integrated supply chain can get higher channel profits. Later, coordinated supply chain with quantity discount strategy is derived and the total profits under the two policies are compared. The numerical example demonstrates that the vendor–buyer coordination along with quantity discount results in an extra total profit and hence it is significant to consider the coordinated vendor–buyer supply chain strategy with quantity discount. Sensitivity analysis is carried out to understand the effect of various key parameters on the optimal solution.

**Keywords** Price-dependent demand · Stock-dependent demand · Deterioration · Supply chain coordination · Quantity discount · All-units quantity discounts

## 8.1 Introduction

Supply chain management can be explained as the systematic coordination of all the business processes like procurement of raw material, selection of vendor, product design, inventory management, manufacturing, and end-customer delivery. Supply chain management has been defined by Lambert et al. (1998) as the coordination of key business processes starting from raw material procurement till end-customer delivery of the product or service in such a way that it adds value to the customers

C. A. Jhaveri (✉) · A. A. Gupta
Institute of Management, Nirma University, Ahmedabad, India
e-mail: chetan@nirmauni.ac.in

A. A. Gupta
e-mail: anuja.gupta@nirmauni.ac.in



N. H. Shah et al. (eds.), *Decision Making in Inventory Management*,
Inventory Optimization, https://doi.org/10.1007/978-981-16-1729-4_8

as well as all the other stakeholders of the organization. A supportive relationship between the buyer and supplier would include mutual trust, sharing information, resource, and profit. This strong relationship is essential so as to have a successful supply chain network (Yang 2004). As a result, a mutually beneficial environment is created between the parties by increasing their joint profits that help the buyer in providing a faster response to the customer demand.

Supply chain coordination is an integral part of an organization, which is used to coordinate and focus on all the relevant resources on the supply chain thus optimizing the use of the available resources and capabilities involved in the overall supply chain. According to Yang (2004), there is rise in the attention given for coordinating the supply chain in organizations due to reasons like depletion in the resources, increase in competition, globalization trend, increasing costs, faster response times, and decreasing product life cycles. Increasing the speed at which materials move in the supply chain would help to reduce the stock level, which would further lead to cost savings for the company.

Nowadays retail stores display a wide array and variety of products of various color, brand, price, and flavor. This is because the companies have observed that a broader collection of products help them to attract more customers into purchasing them. Thus, demand for product is influenced by display stock and does not remain constant. Practically not all the products in the market can have a constant demand, hence it arises the need for development of inventory control models to tackle variable demands. In the past, studies have been done on inventory and pricing strategies for price-dependent demand and supplier's quantity discount schemes. Also, it is observed that product's price as well as its stock-display level affects its demand. It is believed that a large pile of stock display of a particular product in the supermarket will influence customers to purchase it as compared to a product that has a small pile on display.

Retail price of a product has a direct relationship with the demand rate while an inverse relationship with quantity discount price. In order to motivate buyers while making purchasing decisions, often lower costs per unit of goods or materials are offered when purchased in larger quantities. Thus quantity discount is offered by the vendors to persuade buyers into purchasing larger quantities. In the last few years, ecommerce has revolutionized the entire retail industry with the use of quantity discount schemes.

The main goal of this research study is to illustrate the importance of a coordinated supply chain while managing the inventory for deteriorating products having both price as well as stock-dependent demand rate considering the quantity discount scheme of supplier. This has been done by developing a mathematical model for a supply chain system, which is further explained using a numerical illustration to investigate the managerial implication. The second section of this paper contains relevant literature review. Third section includes the mathematical modeling towards the research objective. This section also explains various assumptions and parameters used for modelling. Solution algorithm and a numerical example have been presented in sections four and five respectively. Finally concluding remarks and suggestions for the analyzed model have been provided in the last section.

## 8.2 Literature Review

In this section, various relevant literatures have been discussed and classified based on the type of inventory models.

### *8.2.1 Inventory Models Considering Variable Demand*

Most of the products in the market have a variable demand that is affected by many factors like price, availability, discounts, quality, and stock at display. Thus, there is a need to formulate models based on such factors to manage the inventory so that situations such as over-stocking and under-stocking don't arise. Sarker et al. (1997) have developed a model to achieve the optimal lot-size and order-level for a certain type of goods having varied demand due to decline in quality level. In this model authors have considered two cases wherein they have considered demand to be constant as well as dependent on the stock level. Various other researches have been done where demand depends on the stock, time, or price. Such literatures have been discussed further in this section.

### *8.2.2 Inventory Models Assuming Price Dependent Demand*

For the price-sensitive demand, Li et al. (1996) developed a lot-for-lot joint pricing policy and discussed the benefits obtained as a result of coordination between the buyer and supplier. For the items having linear price function for demand, Wee (1997) came up with an optimal replenishment policy with an objective to maximize the net profit. For the products having constant demand rate, Wee (1998) came up with lot-for-lot discount pricing policy. But neither of these papers considered integrating quantity discount policy with the price-sensitive demand. Qin et al. (2007) developed inventory models with price sensitive demand rate in a coordinated supply chain system. Alfares and Ghaithan (2016) extended the research by Alfares (2015) by considering the price-dependent demand to the existing model.

### *8.2.3 Inventory Models with Stock-Dependent Demand*

Large pile of stock is kept in the display in the supermarket to attract more customers into buying that product mainly because of the variety, visibility, and popularity. Also, it is observed that a low stock display would give out the perception of the product being of low quality or less sold. Thus it can be said that the demand rate for certain types of goods is influenced by the level of stock kept in display in the supermarkets. Stock-dependent consumption rate inventory model was developed by Gupta and Vrat (1986). Their model was anchored on the initial order quantity demand rate instead of the immediate inventory level requirements. Teng and Chang (2005) derived an economic production quantity (EPQ) model to show the dependence of demand rate for specific types of items on selling price per unit and on-display stock with an objective to maximize the profit as well. Goyal and Chang (2009) derived a model to identify the optimal ordering quantity for the buyer as the demand rate depends on the display stock level. Mandal and Phaujdar (1989), Datta and Pal (1990), Urban (2005), Hou and Lin (2006), Chang et al. (2010), Datta and Paul (2001), Sajadieh (2010) have developed and analysed various inventory models considering stock-dependent demand.

### *8.2.4 Deteriorating Products*

As most of the physical products are deteriorating over time, in the recent years, the maintenance of inventories for deteriorating items have received much attention from several researchers. When the utility or usefulness of an item decreases through ways of evaporation, spoilage, or decay; it is known as deterioration of an item. Deterioration may happen during usual period of storage for several products like electronic components, chemicals, drugs, foods, films, etc. Hence, the loss occur due to deterioration of item cannot be ignored. Thus, deterioration of physical goods in the inventory system is a very realistic feature and several researchers realized the necessity to take this fact into consideration while developing inventory models. Giri et al. (1996) developed an inventory model by considering demand for deteriorating items to be stock dependent with a constant rate of deterioration. An objective of this study was to maximize the total profit and find out the appropriate number of orders in the finite planning horizon. Yang and Wee (2000) presented policies for deteriorating items having constant demand rate. Lee and Dye (2012) formulated a deteriorating inventory model having stock-dependent demand. The objective of this model was to know the strategies for optimal replenishments along with maximizing the total profit per unit time. A lot of models for deteriorating items and stock-dependent demand rate in the literature have aimed towards minimizing the inventory costs, but Pando et al. (2018) has considered the rate of deterioration per unit time to be constant part of inventory level with an objective to maximize the total profit per unit time. To study more on deteriorating items literatures can be reviewed from the research done by Raafat (1991), Wee (1999), Yang and Wee (2005) and Sarkar et al. (2013).

### 8.2.5 Supply Chain Coordination

With the increased market competition in the present global markets, organizations are compelled to closely work in collaboration with their suppliers and immediate customers. It is also observed that through better coordination of the supply chain, stocks across the supply chain can be more efficiently managed. In the lack of coordination in the supply chain, each player will act independently to maximize their profit. This may not be the beneficial to the other players of the chain and hence it may result in poor performance of the entire supply chain. The supply chain coordination between the vendor and buyer was first studied by Clark and Scarf (1960); wherein it was assumed that buyer is the sole decision maker of the entire ordering process and hence the solution obtained from such models were not economical for the vendor. Enumerable studies have been done on supply chain coordination. In most cases the resulting profits are distributed equally among supplier and retailer, thus benefiting both the entities. There should be a proper flow of information among the parties in order to have successful supply chain coordination. If one of the parties has better information than others, that might turn out to be his strategic advantage, and might use that information to gain cooperation from other parties. In such cases, the less informed parties try to offer incentives so as to provoke the other party to disclose his private information. The information shared by the parties affects the managers while decision-making. Thus in order to avoid these situations, there should be a mutual flow of information among the parties to maintain the supply chain coordination. Researchers like Goyal and Gupta (1989), Vishwanathan (1998) have come up with inventory models that are applicable to such kind of problems that involve supply chain coordination between vendor and buyer.

### 8.2.6 Inventory Models with Quantity Discount

Researchers recognized that quantity discounts on selling price can provide economic advantages like lower unit purchase cost and lower procurement costs for both vendor and buyer. Some researchers investigated the integrated buyer-vendor inventory problems considering quantity discounts. A fixed order quantity decision model considering the discounting scheme was developed by Lal and Staelin (1984) to benefit the buyers. Vendor oriented optimal quantity discount policy to maximize vendor's profit with no additional cost to the buyer, was studied by many researchers; Monahan (1984) was amongst those early researchers. Monahan's model was generalized and taken further by Lee and Rosenblatt (1986); who developed a fixed order quantity decision model with a discounting scheme that would benefit the buyers. To find out replenishment interval and discount price for any desirable negotiation factor an algorithm was developed by Chakravarty and Martin (1988). Joglekar (1988) has commented on the work done by Monahan (1984) and then explained the model using a numerical illustration. This algorithm was a scheme to build up a mutual cost sharing scheme between buyers and sellers. A simple approach has been proposed by Goyal and Gupta (1990) to identify the optimal order quantity when discounts are offered by the vendor on larger purchases by the buyer. To determine an optimal pricing and replenishment strategy, Weng and Wong (1993) developed a general discount model considering all-unit quantity. For their model Weng and Wong considered demand to be price sensitive. Vendor's quantity discount was considered by Weng (1995) in another study from the point of view of cutting down vendor's operating cost along with increasing buyer's demand. Burwell et al. (1997) developed an inventory model for price-dependent demand considering all-unit quantity discount with an objective to determine the selling price and the optimal lot size. This model by Burwell et al. (1997) was modified by Chang (2013) with an objective to maximize the profit and to determine the accurate optimized values for the lot size and the selling price. Various other inventory models have been developed by Li and Huang (1995), Corbett and Groot (2000), Qi et al. (2004), Li and Liu (2006), Transchel and Minner (2008), Datta and Paul (2001), Zhan et al. (2014), Yin et al. (2015), Alfares and Ghaithan (2016) considering the quantity discount offered by the vendor to the buyer. A manager can use order size-based quantity discounts to achieve channel coordination. Very few inventory models have been developed in the recent literature considering quantity discount scheme. Thus, in this paper, the authors have considered quantity discount as one of the parameters that affects supply chain coordination for deteriorating products while determining the demand rate.

In the literature, several research studies on inventory models were found to be developed for quantity discount and stock-dependent demand while considering supply chain coordination between the vendor and the buyer. There were also models on deterioration, variable demand and price dependent demand, but not a single model has considered all these factors simultaneously. Thus in this research paper, the authors have developed an inventory model for deteriorating products with stock and price dependent demand rate considering quantity discount scheme offered by suppliers to the buyer with the presence of supply chain coordination between the two parties. Table 8.1 summarizes the literatures reviewed for this paper on the basis of various features.

## 8.3 Mathematical Modelling and Analysis

Following assumptions are used to derive the mathematical models in this paper:

(a) The rate of replenishment and lead time are considered to be instantaneous and constant respectively.
(b) The rate of demand decreases linearly with retail price of the product.
(c) All-unit quantity discount is offered by the vendor to the buyer.
(d) The buyer and the vendor share their complete information with each other.
(e) Shortage is not permitted.
(f) A sole unit having a steady deterioration rate is considered.

**Table 8.1** Summary of literature review based on various features

| Authors | Supply chain coordination | Price-dependent demand | Stock-level dependent demand | Quantity discounts | Deterioration |
|---|---|---|---|---|---|
| Alfares (2015) | | | ✓ | ✓ | |
| Alfares and Ghaithan (2016) | | ✓ | | ✓ | |
| Chakravarty and Martin (1988) | | | | ✓ | |
| Chang et al. (2010) | | | ✓ | | ✓ |
| Chang (2013) | | ✓ | | ✓ | |
| Clark and Scarf (1960) | ✓ | | | | |
| Corbett and De Groote (2000) | ✓ | | | ✓ | |
| Datta and Pal (1990) | | | ✓ | | |
| Datta and Paul (2001) | | ✓ | ✓ | | |
| Dye and Yang (2016) | | ✓ | | | ✓ |
| Giri et al. (1996) | | | ✓ | | ✓ |
| Gupta and Vrat (1986) | | | ✓ | | |
| Goyal (1977) | ✓ | | | | |
| Goyal and Gupta (1989) | ✓ | | | | |
| Goyal and Chang (2009) | | | ✓ | | |
| Hou and Lin (2006) | | ✓ | ✓ | | ✓ |
| Joglekar (1988) | | | | ✓ | |
| Lal and Staelin (1984) | | ✓ | | ✓ | |
| Lambert et al. (1998) | ✓ | | | | |

(continued)

**Table 8.1** (continued)

| Authors | Supply chain coordination | Price-dependent demand | Stock-level dependent demand | Quantity discounts | Deterioration |
|---|---|---|---|---|---|
| Lee and Rosenblatt (1986) | | | | ✓ | |
| Lee and Dye (2012) | | | ✓ | | ✓ |
| Li and Huang (1995) | ✓ | | | ✓ | |
| Li et al. (1996) | ✓ | ✓ | | | |
| Li and Liu (2006) | ✓ | | | ✓ | |
| Mandal and Phaujdar (1989) | | | ✓ | | ✓ |
| Monahan (1984) | | | | ✓ | |
| Pando et al. (2018) | | | ✓ | | ✓ |
| Qi et al. (2004) | ✓ | | | ✓ | |
| Qin et al. (2007) | ✓ | ✓ | | ✓ | |
| Raafat (1991) | | | | | ✓ |
| Sajadieh et al. (2010) | ✓ | | ✓ | | |
| Sarkar et al. (2013) | | ✓ | | | ✓ |
| Sarker et al. (1997) | | | ✓ | | ✓ |
| Teng and Chang (2005) | | ✓ | ✓ | | ✓ |
| Transchel and Mirner (2008) | | ✓ | | ✓ | |
| Urban (2005) | | | ✓ | | |
| Viswanathan (1998) | ✓ | | | | |
| Wee (1997) | | ✓ | | | ✓ |
| Wee (1998) | | ✓ | | ✓ | ✓ |
| Wee (1999) | | ✓ | | ✓ | ✓ |

(continued)

**Table 8.1** (continued)

| Authors | Supply chain coordination | Price-dependent demand | Stock-level dependent demand | Quantity discounts | Deterioration |
|---|---|---|---|---|---|
| Weng and Wong (1993) | | ✓ | | ✓ | |
| Weng (1995) | | ✓ | | ✓ | |
| Yang and Wee (2000) | ✓ | | | | ✓ |
| Yang (2004) | | ✓ | | ✓ | ✓ |
| Yang and Wee (2005) | ✓ | | | | ✓ |
| Yin et al. (2015) | ✓ | | | ✓ | |
| Zhang et al. (2014) | ✓ | | | ✓ | |

*Source* own

(g) Deterioration of the units will be considered only after they enter the inventory.
(h) The deteriorated units cannot be repaired or replaced.
(i) Carrying cost will be applied only to the good units.
(j) Supply chain system with single buyer and single vendor is considered.

In this paper, three different cases have been discussed. The vendor–buyer collaboration and quantity discount have not been considered in the first case, while in the second case vendor–buyer integration without quantity discount has been considered. Finally in the third case, buyer-vendor integration as well as quantity discount have been considered simultaneously.

Following parameters related to the vendor are considered for the research:

| | |
|---|---|
| $I_{vi}(t)$ | Level of stock for case $i$, $i = 1, 2, 3$ |
| $C_v$ | Setup cost, $ per cycle |
| $C_{vb}$ | Fixed cost to process each buyer's order |
| $P_v$ | Unit cost for the vendor |
| $F_v$ | Cost of carrying inventory in percentage per year and per dollar |
| $\mathrm{TC}_{vi}$ | Total cost per year for case $i$, $i = 1, 2, 3$ |
| $\mathrm{TP}_{vi}$ | Total profit per year for case $i$, $i = 1, 2, 3$ |
| $S_v$ | Extra profit sharing for case 3 as compared to case 1 ($S_v = \mathrm{TP}_{v3} - \mathrm{TP}_{v1}$) |

Other parameters related to the buyer are as follows:

| | |
|---|---|
| $I_{bi}(t)$ | Level of Inventory for case $i$, $i = 1, 2, 3$ |
| $C_b$ | Buyer's ordering cost, $ per order |
| $P_{bi}$ | Purchase price per unit for the buyer for case $i$, $i = 1, 2$ |
| $F_b$ | Cost of carrying inventory in percentage per year and per dollar |
| $\mathrm{TC}_{bi}$ | Total cost per year for case $i$, $i = 1, 2, 3$ |
| $\mathrm{TP}_{bi}$ | Total profit per year for case $i$, $i = 1, 2, 3$ |
| $S_b$ | Extra profit sharing for case 3 as compared to case 1 ($S_b = \mathrm{TP}_{b3} - \mathrm{TP}_{b1}$) |

Following are the variable parameters:

| | |
|---|---|
| $T_{bi}$ | Replenishment period for the buyer for case $i$, $i = 1, 2, 3$ |
| $n_i$ | Number of replenishments from the vendor to the buyer per cycle for case $i$ |
| $T_{vi}$ | Replenishment period for the vendor for case $i$, $i = 1, 2, 3$ |
| $P_m$ | Retail price for end customer |
| $d$ | Price-dependent demand rate per year |
| $P_{b3}$ | Purchase unit price for the buyer for case 3 |

Other parameters related to buyer and the vendor are as follows:

| | |
|---|---|
| $a$ | Scale parameter for demand rate |
| $b$ | Price–dependent parameter for demand rate |
| $\beta$ | Stock-dependent selling rate parameter |
| $\theta$ | Constant rate of deterioration of on-hand-inventory |
| $\mathrm{TC}_i$ | Total cost per annum ($\mathrm{TC}_{vi}$ and $\mathrm{TC}_{bi}$) for case $i$ |
| $\mathrm{TP}_i$ | Total profit per annum ($\mathrm{TP}_{vi}$ and $\mathrm{TP}_{bi}$) for case $i$ |
| $\gamma$ | Vendor and buyer's extra profit sharing negotiation factor for extra profit sharing |

The inventory level decreases due to the demand and constant deterioration of available stock. Differential equation for inventory system of buyer can be presented as

$$\frac{\mathrm{d}I_{bi}(t)}{\mathrm{d}t} + \theta I_{bi}(t) = -(\alpha + \beta I_{bi}(t)),\ 0 \le t \le T_{bi} \tag{8.1}$$

The boundary condition will take place when $I_{bi}(T_{bi}) = 0$.
The buyer's inventory level using Spiegel (1960) is

$$I_{bi}(t) = \frac{\alpha}{\theta + \beta}\left(e^{(\theta+\beta)(T_{bi}-t)} - 1\right) \tag{8.2}$$

**Case 1**: Supply chain system with the absence of both channel coordination and quantity discount.

The total cost for the system is,

$$TC_{b1} = \text{Buyer's order cost} + \text{Inventory carrying cost} + \text{Buyer's purchasing cost}$$

$$TC_{b1} = \left[ C_b + P_{b1} F_{b1} \int_0^{T_{b1}} I_{b1}(t)dt + P_{b1} I_{b1}(0) \right] / T_{b1}$$

$$TC_{b1} = \frac{\left[ C_b + P_{b1} F_{b1} \left( \frac{\alpha}{(\theta+\beta)^2} \right) \left( e^{(\theta+\beta).T_{b1}} - (\theta+\beta) T_{b1} - 1 \right) + P_{b1} \left( \frac{\alpha}{\theta+\beta} \right) \left( e^{(\theta+\beta).T_{b1}} - 1 \right) \right]}{T_{b1}} \tag{8.3}$$

The three terms in Eq. (8.3) represents cost of ordering, holding cost, and the cost of purchasing, respectively. Using Taylor series approximation, $e^{(\theta+\beta)Tb1}$ in Eq. (8.3) is replaced by $1 + (\theta + \beta)T_{b1} + ½ ((\theta + \beta)T_{b1})^2 + {}^1/_{3!} ((\theta + \beta)T_{b1})^3$, for $(\theta + \beta)T_{b1} << 1$. In Taylor series the fourth term's percentage error is

$$\frac{\frac{(\theta+\beta)^3 T_{b1}^3}{3!}}{1 + (\theta + \beta)T_{b1} + \frac{(\theta+\beta)^2 T_{b1}^2}{2!} + \frac{(\theta+\beta)^3 T_{b1}^3}{3!}}$$

For the small value of $(\theta + \beta)T_{b1}$, the percentage error is very small. It will be even smaller for term higher than four. Hence the term four and onwards are neglected from equation.

The approximated total cost of buyer is,

$$TC_{b1} \cong \left[ \frac{C_b}{T_{b1}} + P_{b1} \times F_{b1} x \frac{\alpha}{2} \times T_{b1} + P_{b1} \times \alpha \left( 1 + \frac{(\theta+\beta)}{2} T_{b1} \right) \right] \tag{8.4}$$

According to the model's assumption; the demand rate has a linearly decreasing function of the retail price while an increasing function of stock-dependent selling rate.

$$d = \alpha + \beta I_{b1}(t) \tag{8.5}$$

where, $\alpha = a - bP_m$.

Buyer's total profit can be calculated by deducting his total cost from his total sales revenue

$$TP_{b1} = (\text{Sales revenue per time unit}) - TC_{b1}$$

Now,

$$\mathrm{SR} = \frac{P_m}{T_{b1}} \int_0^{T_{b1}} (\alpha + \beta I_{b1}(t)\mathrm{d}t)$$

$$\mathrm{SR} = \frac{P_m}{T_{b1}} \left[ \alpha T_{b1} + \frac{\beta\alpha}{(\theta+\beta)^2} \left( e^{(\theta+\beta)(T_{b1})} - (\theta+\beta)T_{b1} - 1 \right) \right]$$

Using Taylor's series approximation;

$$\mathrm{SR} = P_m \times \alpha \left( 1 + \frac{\beta T_{b1}}{2} \right)$$

$$\mathrm{TP}_{b1} = P_m \times \alpha \left( 1 + \frac{\beta T_{b1}}{2} \right) - \mathrm{TC}_{b1} \tag{8.6}$$

We get the following results by taking first derivatives of $\mathrm{TP}_{b1}$ with respect to $T_{b1}$ and $P_m$, and equating these equations to zero.

$$\frac{\partial \mathrm{TP}_{b1}}{\partial T_{b1}} = 0 \tag{8.7}$$

$$\frac{\partial \mathrm{TP}_{b1}}{\partial P_m} = 0 \tag{8.8}$$

The optimal values of $T_{b1}$ and $P_m$ which are denoted by $T_{b1}^*$ and $P_m^*$, will be derived numerically as the solutions obtained in Eqs. (8.7) and (8.8) are not in a closed form.

By using Eqs. (8.4) and (8.5) buyer's optimal total cost is derived for $(\theta+\beta)T_{b1} << 1$ as follows:

$$\mathrm{TC}_{b1}^*(T_{b1}^*, P_m^*) \cong \left[ \frac{C_b}{T_{b1}^*} + P_{b1} \times F_b \times \frac{(a - bP_m^*)}{2} \times T_{b1}^* + P_{b1} \times (a - bP_m^*) \times \left( 1 + \frac{(\theta+\beta)}{2} T_{b1}^* \right) \right] \tag{8.9}$$

The replenishment period for the vendor can be calculated as

$$T_{v1} = n_1 T_{b1}^*, \tag{8.10}$$

where $n_1$ represents the positive integer.

The inventory level for the vendor is

$$I_{v1}(t) = \frac{\alpha}{\theta+\beta} \left[ e^{(\theta+\beta)(n_1 T_{b1}^* - t)} - 1 \right], \tag{8.11}$$

where

$$0 \leq t \leq n_1 T_{b1}^* .$$

As shown in Eq. (8.11) there is an exponential decrease in the inventory level of the vendor. Using Eqs. (8.11) and (8.2), vendor's annual total cost can be derived as follows:

$$\mathrm{TC}_{v1} = \frac{1}{n_1 T_{b1}^*}\left[C_v + n_1 C_{vb} + P_v F_v\left(\int_0^{n_1 T_{b1}^*} I_{v1}(t)\mathrm{d}t - n_1 \int_0^{T_{b1}^*} I_{b1}(t)\mathrm{d}t\right) + P_v I_{v1}(0)\right]$$

$$\mathrm{TC}_{v1} \cong \frac{C_v + n_1 C_{vb}}{n_1 T_{b1}^*} + \frac{P_v F_v \alpha (n_1 - 1) T_{b1}^*}{2} + P_v \alpha\left[1 + \frac{(\theta+\beta)}{2} n_1 T_{b1}^*\right] \quad (8.12)$$

In Eq. (8.12), the first two terms are costs related to the ordering, the next term is saw-tooth shape inventory holding cost while the last term represents costs related to purchasing.

$TP_{v1}$ = (Sales revenue per time unit) − $\mathrm{TC}_{v1}$ Annual total profit for the vendor is

$$\mathrm{TP}_{v1} = \frac{P_{b1} I_{b1}^*(0)}{T_{b1}^*} - \mathrm{TC}_{v1} \approx P_{b1}\alpha\left(1 + \frac{(\theta+\beta)}{2} T_{b1}^*\right) - \mathrm{TC}_{v1} \quad (8.13)$$

Here, $P_{b1}\alpha\left(1 + \frac{(\theta+\beta)}{2} T_{b1}^*\right)$ is the approximated sales revenue for the vendor. Total profit of vendor presented in Eq. (8.13) is a function of a one variable $n_1$. For the vendor's total profit, the optimal policy can be formulated as

$$\text{Maximize } \mathrm{TP}_{v1}(n_1) \text{ for } n = 1, 2, 3, \ldots. \quad (8.14)$$

As $n_1$ is a discrete integer, the following condition must be satisfied for the optimal value of $n_1$, which is denoted by $n_1^*$:

$$\mathrm{TP}_{v1}\left(n_1^* - 1\right) \leq \mathrm{TP}_{v1}\left(n_1^*\right) \geq \mathrm{TP}_{v1}\left(n_1^* + 1\right) \quad (8.2.15)$$

Vendor–buyer system's total profit can be derived using the following equation, when quantity discount and buyer-vendor coordination is not considered

$$\mathrm{TP}_1 = \mathrm{TP}_{b1}\left(T_{b1}^* P_m^*\right) + \mathrm{TP}_{v1}\left(n_1^*\right) \quad (8.16)$$

In case 1, each player makes strategic decisions independently, without considering vendor-buyer coordination. The total annual profit without coordination presented in Eq. (8.16) is a function of multiple decision variables $T_{b1}$, $P_m$ and $n_1$. Buyer first optimizes the decision variables $T_{b1}$ and $P_m$; whereas vendor optimizes the decision variable $n_1$.

**Case 2**: Supply chain system considers channel coordination without vendor's quantity discount

The aim of vendor-buyer coordination is to maximize total channel profit by sharing profit, cost, demand, and stock-related information. This coordination also supports in responding to the customer demand quickly.

Based on Eqs. (8.4) and (8.12), following are the total costs for buyer and vendor, respectively

$$\mathrm{TC}_{b2} = \left[\frac{C_b}{T_{b2}} + P_{b2} \times F_{b2} \times \frac{\alpha}{2} \times T_{b2} + P_{b2} \times \alpha\left(1 + \frac{(\theta+\beta)}{2} T_{b2}\right)\right] \tag{8.17}$$

$$\mathrm{TC}_{v2} = \frac{C_v + n_2 C_{vb}}{n_2 T_{b2}} + \frac{P_v F_v \alpha (n_2 - 1) T_{b2}}{2} + P_v \alpha\left[1 + \frac{(\theta+\beta)}{2} n_2 T_{b2}\right] \tag{8.18}$$

The sum of Eqs. (8.17) and (8.18) represents the coordinated total cost. Based on Eqs. (8.6) and (8.13), following are the profits for buyer and vendor respectively

$$\mathrm{TP}_{b2} = (\text{Sales revenue per time unit}) - \mathrm{TC}_{b2}$$

where,

$$\mathrm{SR} = \frac{P_m}{T_{b2}}\left[\alpha . T_{b2} + \frac{\beta\alpha}{(\theta+\beta)^2}\left(e^{(\theta+\beta)(T_{b2})} - (\theta+\beta)T_{b2} - 1\right)\right]$$

Using Taylor's series approximation, SR can be expressed as,

$$\mathrm{SR} = P_m \times \alpha\left(1 + \frac{\beta T_{b2}}{2}\right)$$

Thus,

$$\mathrm{TP}_{b2} = P_m \times \alpha\left(1 + \frac{\beta T_{b2}}{2}\right) - \mathrm{TC}_{b2} \tag{8.19}$$

$$\mathrm{TP}_{v2} = \frac{P_{b2} I_{b2}(0)}{T_{b2}} - \mathrm{TC}_{v2} \approx P_{b2}\alpha\left(1 + \frac{(\theta+\beta)}{2} T_{b2}\right) - \mathrm{TC}_{v2} \tag{8.20}$$

The total coordinated profit is $\mathrm{TP}_2 = \mathrm{TP}_{b2} + \mathrm{TP}_{v2}$.

Now the objective is to maximize the total coordinated profit,

$$\text{i.e., Max } \mathrm{T}P_2(T_{b2}, P_m, n_2) = \mathrm{TP}_{b2}(T_{b2}, P_m) + \mathrm{TP}_{v2}(n_2) \tag{8.21}$$

In case 2 vendor-buyer coordination is considered. Joint optimization has been done for the three decision variables $T_{b2}$, $P_m$ and $n_2$ rather than optimizing independently as done in case 1.

**Case 3**: Supply chain system when vendor–buyer coordination and quantity discount are considered simultaneously.

In quantity discount scheme, the discount price, $P_{b3}$ is smaller than the unit price, $P_{b1}$ offered in case 1 and 2. Following equation represents the lot size per shipment $Q$ for the buyer:

$$Q = I_{b3}(t = 0) = \frac{\alpha}{\theta + \beta}\left(e^{(\theta+\beta)(T_{b3})} - 1\right) \quad (8.22)$$

The delivery quantity from vendor to the buyer per annum can be derived as follows:

$$\frac{I_{b3}(0)}{T_{b3}} = \frac{\alpha}{T_{b3}(\theta + \beta)}\left(e^{(\theta+\beta)(T_{b3})} - 1\right) \approx d\left(1 + \frac{(\theta + \beta)}{2} T_{b3}\right) \quad (8.23)$$

Likewise, following are the annual total cost for buyer and vendor respectively:

$$\mathrm{TC}_{b3} \cong \left[\frac{C_b}{T_{b3}} + P_{b3} \times F_{b3}\, x\, \frac{\alpha}{2} \times T_{b3} + P_{b3} \times \alpha\left(1 + \frac{(\theta + \beta)}{2} T_{b3}\right)\right] \quad (8.24)$$

$$\mathrm{TC}_{v3} = \frac{C_v + n_3 C_{vb}}{n_3 T_{b3}} + \frac{P_v F_v \alpha (n_3 - 1) T_{b3}}{2} + P_v \alpha\left[1 + \frac{(\theta + \beta)}{2} n_3 T_{b3}\right] + (P_{b1} - P_{b3})\left(1 + \frac{(\theta + \beta)}{2} T_{b3}\right) \quad (8.25)$$

When the vendor offers a quantity discount, there is an additional cost which is shown as the last term in Eq. (8.25). Following is the total profit of buyer and vendor respectively:

$$\mathrm{TP}_{b3} = \text{Sales revenue per time unit(SR)} - \mathrm{TC}_{b3} \quad (8.26)$$

$$\mathrm{TP}_{v3} = \frac{P_{b3} I_{b3}(0)}{T_{b3}} - TC_{v3} = P_{b3}.\alpha\left(1 + \frac{(\theta + \beta)}{2} T_{b3}\right) - \mathrm{TC}_{v3} \quad (8.27)$$

The difference between $\mathrm{TP}_{b3}$ and $\mathrm{TP}_{b1}$ is the buyer's extra profit, denoted by $S_b$ is shown below:

$$S_b = \mathrm{TP}_{b3} - \mathrm{TP}_{b1} \quad (8.28)$$

The vendor's extra profit is the difference between $\mathrm{TP}_{v3}$ and $\mathrm{TP}_{v1}$, defined by $S_v$.

$$S_v = \mathrm{TP}_{v3} - \mathrm{TP}_{v1} \quad (8.29)$$

The uncoordinated total profit in case 1 ($TP_1$) is less than the coordinated total profit in case 2 ($TP_2$), also the coordinated total profit in case 3 ($TP_3$) is greater than that of $TP_2$; hence it can be said that $TP_3$ is greater than $TP_1$. This relationship between the total profit of case 3 and case 1 for both the vendor and the buyer, denoted as $S_b$ and $S_v$ is defined as:

$$S_v = \gamma S_b, \gamma \geq 0, \tag{8.30}$$

where, $\gamma =$ negotiation factor.

When the negotiation factor $\gamma = 0$, all the extra profit is given to the buyer. When $\gamma = 1$, all the extra profit is distributed equally between the buyer and vendor. While if $\gamma > 1$, all extra profit is given to the vendor. Following is the optimization problem for case 3:

$$\text{Maximize } TP_3(T_{b3}, P_m, n_3) = TP_{b3}(T_{b3}, P_m) + TP_{v3}(n_3) \tag{8.31}$$

Here, $TP_3$ is the function of the three variables $n_3$, $T_{b3}$ and $P_m$ .

## 8.4 Solution Procedure

For case 1, value of $n_1$ is to be determined such that $TP_1$ presented as Eq. (8.16) can be maximized. Here $T_{b1}$ and $P_m$ are optimized by buyer first and then variable $n_1$ is optimized by the vendor such that Eqs. (8.14) and (8.15) are satisfied.

For case 2, value of $n_2$ is to be determined such that $TP_2$ (8.21) can be maximized. Following procedure can be used to derive $n_2$ i.e. the number of delivery per order, as it is a discrete variable:

(a) Given a range of $n_2$ values, first with respect to $P_m$ and $T_{b2}$ obtain the partial derivative of $TP_2$ and equate them to zero; for a given range of $n_2$ values. For each $n_2$, $P_m(n_2)$ and $T_{b2}(n_2)$ are the optimal value of $Pm$ and $T_{b2}$ respectively.

(b) Derive ${n_2}^*$, the optimal value of $n_2$, such that

$$\begin{aligned} TP_2\left(T_{b2}\left(n_2^* - 1\right),\ n_2^* - 1,\ P_m\left(n_2^* - 1\right)\right) &\leq TP_2\left(T_{b2}\left(n_2^*\right),\ n_2^*,\ P_m\left(n_2^*\right)\right) \\ &\geq TP_2\left(T_{b2}\left(n_2^* + 1\right),\ n_2^* + 1,\ P_m\left(n_2^* + 1\right)\right) \end{aligned}$$

For case 3, Eq. (8.31) has to be maximized to determine the value of decision variable $n_3$. In order to maximize the total profit $TP_3$; find partial derivatives of $TP_3$ with respect to $T_{b3}$ and $P_m$ need to be set equal to zero as shown below:

$$\frac{\partial TP_3}{\partial T_{b3}} = 0 \tag{8.32}$$

$$\frac{\partial \mathrm{TP}_3}{\partial P_m} = 0 \tag{8.33}$$

In case 3, quantity discount is offered to the buyer, thus solution procedure in case 3 is different than that in case 2. While applying the procedure, the solution obtained from Eqs. (8.31) to (8.33) must be rounded up. The values of $P_m$, $T_{b3}$ and $\mathrm{TP}_3(P_m, T_{b3})$ should be rounded to the nearest two decimals, while the order quantity, $Q$ should be rounded to the nearest integer.

For case 3, Eq. (8.31) has to be maximized to determine the value of $n_3$. Following procedure will be used to derive the value of $n_3$ in case 3. Given a range of $n_3$ values, first find the partial derivative of $\mathrm{TP}_3$ with respect to $T_{b3}$ and $P_m$. Equate these equations to zero and solve to get the value of $T_{b3}$ and $P_m$.

Step 1: For a given range of $n_3$ values, optimal values of $T_{b3}$ and $P_m$ can be obtained using the following procedure:

a. Put $\mathrm{TP}_{3\max} = 0$ and $j = J$
b. Solve for $P_m$ and $T_{b3}$ after replacing all the given values $(a, b, \beta, \theta,)$ and $P_{b3} = c_j$ in Eqs. (8.32) and (8.33). Obtain order quantity $Q$ from Eq. (8.22).
   The obtained solution will be feasible if $Q$ lies in the correct purchase cost range i.e. $q_{j-1} \le Q < q_j$. To calculate $\mathrm{TP}_3(P_m, T_{b3})$ put the optimal values of $T_{b3}$ and $P_m$ in Eq. (8.31). Set $\mathrm{TP}_{3\max} = \mathrm{TP}_3(P_m, T_{b3})$ if $\mathrm{TP}_3(P_m, T_{b3}) > \mathrm{TP}_{3\max}$. Next, go to step (e).
   The obtained solution is not feasible if order quantity $Q$ does not fall in the right purchasing cost range. In that case, follow step (c).
c. Since value of $Q$ is obtained in step (b) does not fall in the range $q_{j-1} \le Q < q_j$, it is not a feasible quantity. To take advantage of price discount the order quantity must be at price break i.e. $Q = q_{j-1}$. Substitute this value of Q in the equation of $P_m$ (see appendix).
   Solve for $T_{b3}$ by substituting $Q = q_{j-1}$, $P_{b3} = c_j$ and other given values $(a, b, \beta, \theta)$ along with $P_m$ into Eq. (8.32). To calculate $\mathrm{TP}_{3j}(P_m, T_{b3})$ put the values of $Q = q_{j-1}$ and the corresponding values of $T_{b3}$ and $P_m$ obtained above into Eq. (8.31). Set $\mathrm{TP}_{3\max} = \mathrm{TP}_{3j}(P_m, T_{b3})$ if $\mathrm{TP}_{3j} > \mathrm{TP}_{3\max}$. Go to step (d).
d. Set $j = j - 1$ if $j \ge 2$ and go to step (b).
   Follow step (e) if $j = 1$.
e. The obtained solution is the feasible solution associated with $\mathrm{TP}_{3\max}$. By specifying the optimal values of $T_{b3}$, $P_m$, $\mathrm{TP}_{3j}(P_m, T_{b3})$, the obtained solution can be defined for a given value of $n_3$. This ends the process.

Step 2:

$n_3^*$ is the optimal value of $n_3$ which can be derived by satisfying following condition:

$$\begin{aligned}\mathrm{TP}_3\left(T_{b3}\left(n_3^* - 1\right), n_3^* - 1, P_m\left(n_3^* - 1\right)\right) &\le \mathrm{TP}_3\left(T_{b3}\left(n_3^*\right), n_3^*, P_m\left(n_3^*\right)\right) \\ &\ge \mathrm{TP}_3\left(T_{b3}\left(n_3^* + 1\right), n_3^* + 1, P_m\left(n_3^* + 1\right)\right)\end{aligned} \tag{8.34}$$

## 8.5 Numerical Example

The solution procedure discussed in the previous section can be explained through the following numerical example. Data which are considered to illustrate the derived model and the proposed algorithm are as follows:

Scale parameter, $a = 2000$.
Price-dependent parameter, $b = 33$.
Stock-dependent selling rate parameter, $\beta = 0.03$.
Carrying cost for vendor, in percentage per annum per dollar, $F_v = 0.2$
Setup cost for vendor, $C_v = \$6000$.
Fixed cost for vendor to process each order placed by buyer, $C_{vb} = \$100$.
Unit cost for vendor, $P_v = \$20$.
Carrying cost for buyer, in percentage per annum per dollar, $F_b = 0.2$
Buyer's ordering cost, $C_b = \$100$.
Purchased unit price for buyer without price discount, $P_{b1} = P_{b2} = \$33$.
Deterioration rate, $\theta = 0.05$.
Negotiation factor, $\gamma = 0$ or 1.

As per the model assumption, all-unit discount scheme is being offered by the vendor to the buyer wherein the buyer gets discount based on the quantity purchased by him.

Following is the price range, based on which per unit cost for the buyer can be determined:

| No. of units | | Cost per unit | |
|---|---|---|---|
| $q_1$ | 0–299 | $c_1$ | 33 |
| $q_2$ | 300–599 | $c_2$ | 31.5 |
| $q_3$ | More than 599 | $c_3$ | 30 |

The computational results are presented in Table 8.2. The annual demand, buyer's unit purchase price and replenishment period, number of replenishments from vendor, the optimum retail price of product, and associated total annual profit for buyer and vendor for all the three cases are presented in Table 8.2.

The number of replenishments for case 1, i.e,. supply chain without integration is $n = 9$; the associated retail price and buyer's replenishment period are \$47.30 and 0.2413 years are also shown in Table 8.2. The corresponding annual demand for the product is 441 units. The total annual profit for buyer and vendor are \$5450 and \$213 respectively. The total annual profit for the supply chain without integration is \$ 5663.

For case 2, when supply chain coordination is considered, the optimal values of the decision parameters retail price and buyer's replenishment period are \$43.34 and 0.6533 years. The number of replenishment from vendor to buyer '$n$' is 3 and the annual demand of the product is 575 units. The total annual profit for buyer and vendor are \$4262 and \$2300 respectively. The optimal value of coordinated channel's total annual profit is \$6562. The total annual profit for the coordinated channel is \$899 is higher than the total profit of supply chain without coordination. Due to channel coordination vendor profit is increased from \$213 to \$2300 whereas buyers profit is declined from \$5450 to \$4262. Since coordination in the supply chain is beneficial to vendor only, buyer would not like to participate in the coordinated strategy and resist to share the information.

To encourage the buyer to participate in the channel coordination, vendor may offer quantity discount and can share profit benefit with the buyer; which is earned due to coordination strategy. When supply chain coordination and quantity discount are considered simultaneously, the channel's annual total profit is increased to \$6629 with the optimal unit discounted purchase price of \$31.50. The percentage of extra total profit ($PETP_3$) is 17.06% which is higher than 15.87%, the percentage of extra total profit ($PETP_2$) when coordination is considered without discount policy.

From Table 8.2, it can be observed that the vendor can earn greater profits by the adoption of an appropriate discount strategy. The increase in the channel annual total profit from case 1 to case 3 is \$ 966 (\$6629–\$5663). Due to supply chain coordination, in case 2 and case 3; vendor's extra profit $S_v$ is increased by \$2087 and \$1826, whereas buyer's extra profit Sb is negative as profit is decreased by \$1188 and \$860 respectively. If all extra profit earned in case 3 is offered to the buyer (i.e. negotiation factor $\gamma = 0$), then buyer and vendor's annual total profit will be \$5556 and \$1073, which is higher than case 1, where coordination and quantity discount are not considered in supply chain system. Adoption of coordination along with quantity discount policy is beneficial to both vendor and buyer.

The numerical results obtained through the above solution procedure shows that $TP_n$ is strictly concave in $T_b$ and $P_m$ (Fig. 8.1). Hence, the local maximum value of objective function obtained here from proposed solution procedure is indeed the global maximum solution.

### 8.5.1 Sensitivity Analysis

The relative impact of various parameters on the optimal solution obtained in case 3 is studied through sensitivity analysis. The sensitivity analysis is performed by changing value of each given parameters by −20%, −10%, +10%, and +20%, taking one parameter at a time and keeping the value of other parameters unchanged. The results of the sensitivity analysis are given in Tables 8.3, 8.4, 8.5, 8.6 and 8.7. The results of the sensitivity analysis show the impact of changes in the key parameters on the decision variables $P_m$, $n_3$, $d$, $P_{b3}$, $T_{b3}$, $TP_1$, $TP_2$ and $TP_3$.

From the results shown in Table 8.3, it is observed that $PETP_3$ changes significantly in the range 9% to 49%, when the price-sensitive parameter $b$ changes. The change in $b$ and $PETP_3$ is positively correlated. This indicates that when b increases, it is more significant to consider coordination strategy with price discount (Fig. 8.2).

It can be observed from Tables 8.4 and 8.5, when $C_v$, $C_b$ and $C_{vb}$ increases, total annual profit decrease but PETP3 increases. Hence, it is very important to take into

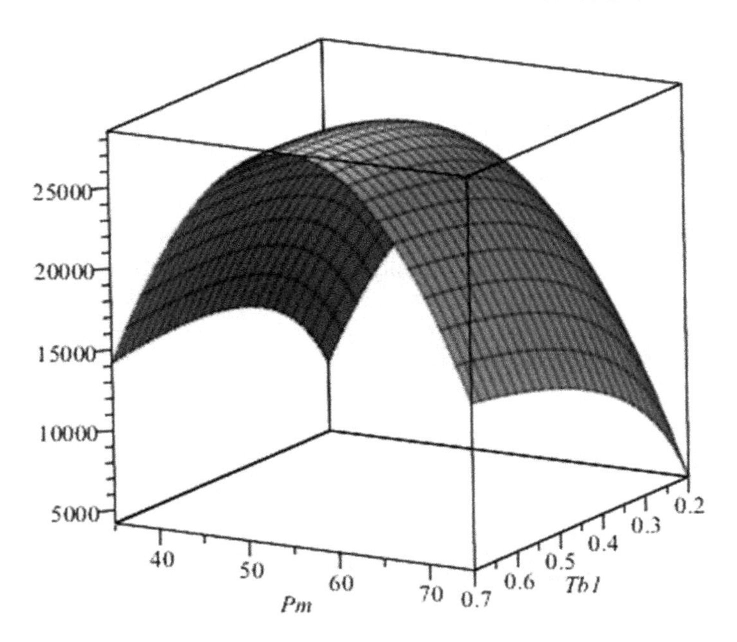


**Fig. 8.1** Concavity of total profit function. *Source* own

**Table 8.2** The optimal solution at various cases when $\theta = 0.05$

| Case $i$ | $i = 1$ | $i = 2$ | $i = 3$ |
|---|---|---|---|
| $P_m$ | 47.30 | 43.34 | 43.35 |
| $d$ | 441 | 575 | 578 |
| $P_{bi}$ | 33 | 33 | 31.50 |
| $n_i$ | 9 | 3 | 2 |
| $T_{bi}$ | 0.2413 | 0.6533 | 0.9599 |
| $TP_{bi}$ | 5450 | 4262 | 4590 |
| $TP_{vi}$ | 213 | 2300 | 2039 |
| $TP_i$ | 5663 | 6562 | 6629 |
| $PETP_i$ | – | 0.1587 | 0.1706 |

$PETP_i$: Percentage of extra total profit for case $i$ compared to case 1;
$PETP_i = (TP_i - TP_1)/TP_1$
*Source* own

account both the integration and the quantity discount when the costs related to order processing for the player of supply chain increase.

From Table 8.6, we can see that total annual profit decreases significantly, and PETP3 increases when rate of deterioration θ increases. This shows that supply chain coordination with quantity discount strategy is advisable when rate of deterioration increases over the time period.

Table 8.3 Sensitivity analysis for price-dependent parameter $b$

| $b$ | $P_m$ | $d$ | $P_{b3}$ | $n_3$ | $T_{b3}$ | $TP_1$ | $TP_2$ ($PETP_2$) | | $TP_3$ ($PETP_3$) | |
|---|---|---|---|---|---|---|---|---|---|---|
| 26.4 | 50.70 | 671 | 30 | 2 | 0.9103 | 12,177 | 13,099 | (7.57%) | 13,276 | (9.03%) |
| 29.7 | 46.58 | 625 | 30 | 2 | 0.9378 | 8487 | 9400 | (10.78%) | 9564 | (12.69%) |
| 33 | 43.35 | 578 | 31.5 | 2 | 0.9599 | 5663 | 6562 | (15.87%) | 6629 | (17.06%) |
| 36.3 | 40.72 | 529 | 31.5 | 2 | 0.9995 | 3479 | 4352 | (25.09%) | 4409 | (26.73%) |
| 39.6 | 38.54 | 479 | 31.5 | 3 | 0.7195 | 1786 | 2617 | (46.53%) | 2668 | (49.38%) |

*Source* own

The results shown in Table 8.7 indicate that with increase in the stock-dependent selling rate parameter $\beta$, PETP3 increases significantly. Hence, it is preferable to adopt coordination with discount policy when stock-dependent selling rate parameter $\beta$ increases.

As price-sensitive parameter $b$, rate of deterioration $\theta$ and $C_V$ increases, demand decreases significantly whereas if stock-dependent selling rate parameter $\beta$, $C_{vb}$ and $C_b$ increases, demand also increases. The effect of stock-dependent selling rate parameter $\beta$ is more significant on $T_{b3}$ and $n$. Retail price of product is more sensitive to price-sensitive parameter $b$, $C_{vb}$ and $C_b$ as compared to other parameters (Figs. 8.3 and 8.4).

## 8.6 Conclusions

This study presents coordinated supply chain system with variable demand rate and a variable unit purchase cost. In this study, more realistic model parameters like stock-dependent selling rate and deterioration are considered in deriving the model. A model has been derived, and an efficient solution procedure has been discussed to determine the optimal unit retail selling price and replenishment cycle. The impact of price-sensitive parameter b, deterioration, stock-dependent selling rate on total annual profit, demand of product, and retail selling price are reported. The results indicate that supply chain coordination with quantity discount increases the extra total profit gain of about 17.06%.

Supply chain coordination helps in optimizing the overall system rather than its individual players and not only increases total annual profits but also reduce variability in demand and inventory level, resulting in more efficient supply chain. The result of sensitivity analysis shows that the effects of price-sensitive parameter, stock-dependent selling rate and deterioration on the total annual profit are very significant, and hence cannot be ignored while deriving the supply chain model.

**Table 8.4** Sensitivity analysis for $C_V$

| $C_v$ | $P_m$ | $d$ | $P_{b3}$ | $n_3$ | $T_{v3}$ | $T_{b3}$ | $TP_1$ | $TP_2$ ($PETP_2$) | | $TP_3$ ($PETP_3$) | |
|---|---|---|---|---|---|---|---|---|---|---|---|
| 4800 | 43.13 | 591 | 30 | 1 | 1.6584 | 1.6584 | 6236 | 7215 | (15.70%) | 7309 | (17.2065%) |
| 5400 | 43.20 | 583 | 31.5 | 2 | 1.8190 | 0.9095 | 5940 | 6877 | (15.77%) | 6950 | (17.0034%) |
| 6000 | 43.35 | 578 | 31.5 | 2 | 1.9198 | 0.9599 | 5663 | 6562 | (15.87%) | 6629 | (17.0581%) |
| 6600 | 43.47 | 574 | 30 | 2 | 2.0400 | 1.0200 | 5394 | 6264 | (16.13%) | 6410 | (18.8357%) |
| 7200 | 43.61 | 570 | 30 | 2 | 2.1350 | 1.0675 | 5146 | 5978 | (16.17%) | 6122 | (18.9662%) |

$T_{vi} = T_{bi} \times n_i$

*Source* own

**Table 8.5** Sensitivity analysis for $C_{vb}$ and $C_b$

| $C_{vb}$, $C_b$ | $P_m$ | $d$ | $P_{b3}$ | $n_3$ | $T_{b3}$ | $TP_1$ | $TP_2$ ($PETP_2$) | | $TP_3$ ($PETP_3$) | |
|---|---|---|---|---|---|---|---|---|---|---|
| 80, 80 | 43.29 | 577 | 31.5 | 3 | 0.6516 | 5769 | 6624 | (14.82%) | 6680 | (15.79%) |
| 90, 90 | 43.34 | 578 | 31.5 | 2 | 0.9566 | 5712 | 6593 | (15.42%) | 6650 | (16.42%) |
| 100, 100 | 43.35 | 578 | 31.5 | 2 | 0.9599 | 5663 | 6562 | (15.87%) | 6629 | (17.06%) |
| 110, 110 | 41.80 | 629 | 30 | 2 | 0.9319 | 5613 | 6532 | (16.37%) | 6619 | (17.92%) |
| 120, 120 | 41.93 | 625 | 30 | 2 | 0.9383 | 5564 | 6505 | (16.91%) | 6610 | (18.80%) |

*Source* own

In this study, the problem of simultaneously determining a pricing and ordering strategy for deteriorating product is addressed. The models can be applied for efficient supplier management in system like super market and stationery stores to determine optimal ordering and pricing policy. Retailer can use this model to optimize this retail unit price and inventory control variables.

The above model can be extended by considering different form of demand rate like nonlinear function of inventory level or retail price. Also, consideration of shortages, permissible delay in payment in the model can help to extend the model further. Additionally, this model can be extended further for deteriorating product with a two-parameter Weibull distribution.

**Table 8.6** Sensitivity analysis for $\theta$

| $\theta$ | $P_m$ | $d$ | $P_{b3}$ | $n_3$ | $T_{v3}$ | $T_{b3}$ | $\text{TP}_1$ | $\text{TP}_2$ ($\text{PETP}_2$) | | $\text{TP}_3$ ($\text{PETP}_3$) | |
|---|---|---|---|---|---|---|---|---|---|---|---|
| 0.01 | 43.11 | 586 | 30 | 2 | 2.0742 | 1.0371 | 6144 | 7026 | (14.36%) | 7172 | (16.73%) |
| 0.04 | 42.20 | 579 | 30 | 2 | 1.9114 | 0.9557 | 5780 | 6675 | (15.48%) | 6788 | (17.44%) |
| 0.045 | 41.93 | 578 | 30 | 2 | 1.8808 | 0.9404 | 5722 | 6618 | (15.66%) | 6710 | (17.27%) |
| 0.05 | 43.35 | 578 | 31.5 | 2 | 1.9198 | 0.9599 | 5663 | 6562 | (15.87%) | 6629 | (17.06%) |
| 0.055 | 43.50 | 580 | 30 | 1 | 1.8535 | 1.8535 | 5606 | 6507 | (16.07%) | 6575 | (17.29%) |
| 0.06 | 43.53 | 579 | 30 | 1 | 1.8406 | 1.8406 | 5549 | 6452 | (16.27%) | 6523 | (17.55%) |
| 0.10 | 43.71 | 572 | 30 | 1 | 1.7466 | 1.7466 | 5110 | 6030 | (18.00%) | 6121 | (19.78%) |

*Source* own

**Table 8.7** Sensitivity analysis for stock-dependent selling rate parameter $\beta$

| $\beta$ | $P_m$ | $d$ | $P_{b3}$ | $n_3$ | $T_{b3}$ | $TP_1$ | $TP_2$ ($PETP_2$) | | $TP_3$ ($PETP_3$) | |
|---|---|---|---|---|---|---|---|---|---|---|
| 0.024 | 43.2902 | 576 | 31.5 | 3 | 0.6595 | 5701 | 6581 | (15.44%) | 6637 | (16.42%) |
| 0.027 | 43.3039 | 576 | 31.5 | 3 | 0.6588 | 5682 | 6572 | (15.66%) | 6628 | (16.65%) |
| 0.03 | 43.35 | 578 | 31.5 | 2 | 0.9599 | 5663 | 6562 | (15.87%) | 6629 | (17.06%) |
| 0.033 | 43.49 | 582 | 30 | 1 | 1.8780 | 5644 | 6553 | (16.11%) | 6665 | (18.09%) |
| 0.036 | 43.50 | 584 | 30 | 1 | 1.8893 | 5625 | 6552 | (16.48%) | 6702 | (19.15%) |
| 0.05 | 43.5491 | 590 | 30 | 1 | 1.9451 | 5533 | 6566 | (18.67%) | 6880 | (24.34%) |

*Source* own

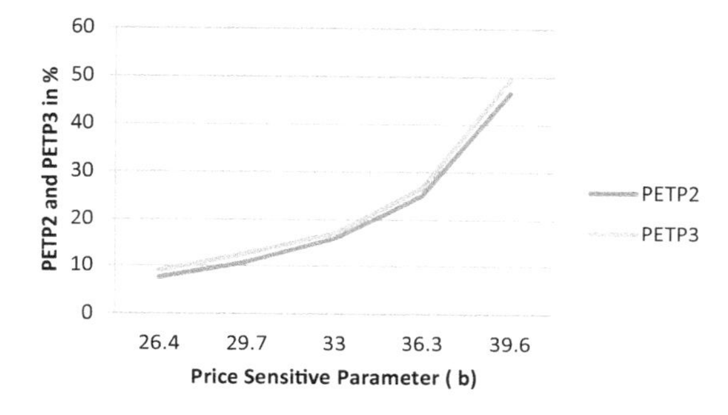


**Fig. 8.2** $PETP_2$ and $PETP_3$ versus price-sensitive parameter. *Source* own

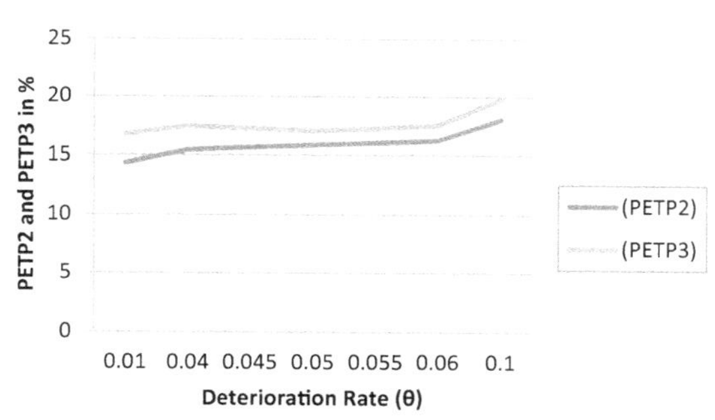


**Fig. 8.3** $PETP_2$ and $PETP_3$ versus deterioration rate. *Source* own

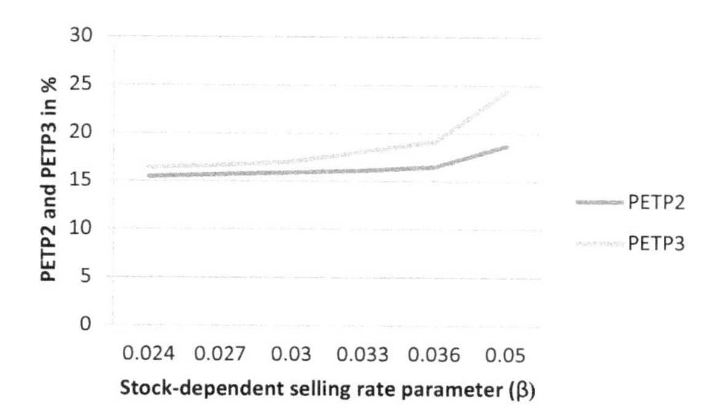


**Fig. 8.4** $PETP_2$ and $PETP_3$ versus Stock-dependent selling rate. *Source* own

## Appendix

From Eq. (8.22), $P_m$ can be expressed as a function at $Q$ and $T_{b3}$ as follows:

$Q = I_{mb} = (a - b.P_m)T_{b3}[1 + (\theta + \beta)T_{b3}]$ (Using Taylor series approximation)

Using above equation, $P_m$ can be expressed as:

$$P_m = \frac{1}{b}\left[a - \frac{Q}{T_{b3}\left(1 + \frac{(\theta+\beta)T_{b3}}{2}\right)}\right]$$

## References


Alfares HK (2015) Maximum-profit inventory model with stock-dependent demand, time-dependent holding cost, and all-units quantity discounts. Math Model Anal 20(6):715–736

Alfares H, Ghaithan A (2016) Inventory and pricing model with price-dependent demand, time-varying holding cost, and quantity discounts. Comput Ind Eng 94:170–177. https://doi.org/10.1016/j.cie.2016.02.009

Burwell TH, Dave DS, Fitzpatrick KE, Roy MR (1997) Economic lot size model for price-dependent demand under quantity and freight discounts. Int J Prod Econ 48(2):141–155

Chakravarty A, Martin G (1988) An optimal joint buyer-seller discount pricing model. Comput Oper Res 15(3):271–281. https://doi.org/10.1016/0305-0548(88)90040-8

Chang CT, Teng JT, Goyal SK (2010) Optimal replenishment policies for non-instantaneous deteriorating items with stock-dependent demand. Int J Prod Econ 123(1):62–68

Chang HC (2013) A note on an economic lot size model for price-dependent demand under quantity and freight discounts. Int J Prod Econ 144(1):175–179. https://doi.org/10.1016/j.ijpe.2013.02.001

Clark AJ, Scarf H (1960) Optimal policies for a multi-echelon inventory problem. Manage Sci 6(4):475–490

Corbett CJ, De Groote X (2000) A supplier's optimal quantity discount policy under asymmetric information. Manage Sci 46(3):444–450

Datta TK, Pal AK (1990) A note on an inventory model with inventory-level-dependent demand rate. J Oper Res Soc 41(10):971–975

Datta TK, Paul K (2001) An inventory system with stock-dependent, price-sensitive demand rate. Prod Plan Control 12(1):13–20

Dye CY, Yang CT (2016) Optimal dynamic pricing and preservation technology investment for deteriorating products with reference price effects. Omega 62:52–67

Giri B, Pal S, Goswami A, Chaudhuri K (1996) An inventory model for deteriorating items with stock-dependent demand rate. Eur J Oper Res 95(3):604–610. https://doi.org/10.1016/0377-2217(95)00309-6

Goyal SK (1977) An integrated inventory model for a single supplier-single customer problem. Int J Prod Res 15(1):107–111. https://doi.org/10.1080/00207547708943107

Goyal SK, Gupta YP (1989) Integrated inventory models: the buyer-vendor coordination. Eur J Oper Res 41(3):261–269. https://doi.org/10.1016/0377-2217(89)90247-6

Goyal SK, Gupta OK (1990) A simple approach to the discount purchase problem. J Oper Res Soc 41(12):1169–1170

Goyal SK, Chang CT (2009) Optimal ordering and transfer policy for an inventory with stock dependent demand. Eur J Oper Res 196(1):177–185
Gupta R, Vrat P (1986) Inventory model for stock-dependent consumption rate. Opsearch 23(1):19–24
Hou KL, Lin LC (2006) An EOQ model for deteriorating items with price-and stock-dependent selling rates under inflation and time value of money. Int J Syst Sci 37(15):1131–1139
Joglekar PN (1988) Note—comments on "A quantity discount pricing model to increase vendor profits." Manage Sci 34(11):1391–1398
Lal R, Staelin R (1984) An approach for developing an optimal discount pricing policy. Manage Sci 30(12):1524–1539. https://doi.org/10.1287/mnsc.30.12.1524
Lambert DM, Cooper MC, Pagh JD (1998) Supply chain management: implementation issues and research opportunities. Int J Logist Manage 9(2):1–20
Lee HL, Rosenblatt MJ (1986) A generalized quantity discount pricing model to increase supplier's profits. Manage Sci 32(9):1177–1185. https://doi.org/10.1287/mnsc.32.9.1177
Lee Y, Dye C (2012) An inventory model for deteriorating items under stock-dependent demand and controllable deterioration rate. Comput Ind Eng 63(2):474–482. https://doi.org/10.1016/j.cie.2012.04.006
Li SX, Huang Z (1995) Managing buyer-seller system cooperation with quantity discount considerations. Comput Oper Res 22(9):947–958
Li SX, Huang Z, Ashley A (1996) Inventory, channel coordination and bargaining in a manufacturer-retailer system. Ann Oper Res 68(1):47–60
Li J, Liu L (2006) Supply chain coordination with quantity discount policy. Int J Prod Econ 101(1):89–98
Mandal BA, Phaujdar S (1989) An inventory model for deteriorating items and stock-dependent consumption rate. J Oper Res Soc 40(5):483–488
Monahan J (1984) A quantity discount pricing model to increase vendor profits. Manage Sci 30(6):720–726. https://doi.org/10.1287/mnsc.30.6.720
Pando V, San-José L, García-Laguna J, Sicilia J (2018) Optimal lot-size policy for deteriorating items with stock-dependent demand considering profit maximization. Comput Industr Eng 117:81–93. https://doi.org/10.1016/j.cie.2018.01.008
Qi X, Bard J, Yu G (2004) Supply chain coordination with demand disruptions. Omega 32(4):301–312. https://doi.org/10.1016/j.omega.2003.12.002
Qin Y, Tang H, Guo C (2007) Channel coordination and volume discounts with price-sensitive demand. Int J Prod Econ 105(1):43–53
Raafat F (1991) Survey of literature on continuously deteriorating inventory models. J Oper Res Soc 42(1):27–37
Sajadieh MS, Thorstenson A, Jokar MRA (2010) An integrated vendor–buyer model with stock-dependent demand. Transp Res Part E: Logist Transp Rev 46(6):963–974
Sarkar B, Saren S, Wee HM (2013) An inventory model with variable demand, component cost and selling price for deteriorating items. Econ Model 30:306–310. https://doi.org/10.1016/j.econmod.2012.09.002
Sarker BR, Mukherjee S, Balan CV (1997) An order-level lot size inventory model with inventory-dependent demand and deterioration. Int J Prod Econ 48:227–236
Spiegel M (1960) Applied differential equations. Prentice, Englewood Cliffs
Teng J, Chang C (2005) Economic production quantity models for deteriorating items with price- and stock-dependent demand. Comput Oper Res 32(2):297–308. https://doi.org/10.1016/s0305-0548(03)00237-5
Transchel S, Mirner S (2008) Coordinated lot-sizing and dynamic pricing under a supplier all-units quantity discount. Bus Res 1(1):125–141. https://doi.org/10.1007/bf03342706
Urban TL (2005) Inventory models with inventory-level-dependent demand: a comprehensive review and unifying theory. Eur J Oper Res 162(3):792–804
Viswanathan S (1998) Optimal strategy for the integrated vendor-buyer inventory model. Eur J Oper Res 105(1):38–42

Wee HM (1997) A replenishment policy for items with a price-dependent demand and a varying rate of deterioration. Prod Plan Control 8(5):494–499. https://doi.org/10.1080/095372897235073

Wee HM (1998) Optimal buyer-seller discount pricing and ordering policy for deteriorating items. Eng Econ 43(2):151–168

Wee H (1999) Deteriorating inventory model with quantity discount, pricing and partial backordering. Int J Prod Econ 59(1–3):511–518. https://doi.org/10.1016/s0925-5273(98)00113-3

Weng KZ, Wong RT (1993) General models for the supplier's all-unit quantity discount policy. Nav Res Logist 40(7):971–991. https://doi.org/10.1002/1520-6750(199312)40:7%3c971::aid-nav3220400708%3e3.0.co;2-t

Weng KZ (1995) Modeling quantity discounts under general price-sensitive demand functions: optimal policies and relationships. Eur J Oper Res 86(2):300–314. https://doi.org/10.1016/0377-2217(94)00104-k

Yang PC, Wee HM (2000) Economic ordering policy of deteriorated item for vendor and buyer: an integrated approach. Prod Plan Control 11(5):474–480

Yang P (2004) Pricing strategy for deteriorating items using quantity discount when demand is price sensitive. Eur J Oper Res 157(2):389–397. https://doi.org/10.1016/s0377-2217(03)00241-8

Yang PC, Wee HM (2005) A win-win strategy for an integrated vendor-buyer deteriorating inventory system. Math Model Anal, 541–546

Yin S, Nishi T, Grossmann IE (2015) Optimal quantity discount coordination for supply chain optimization with one manufacturer and multiple suppliers under demand uncertainty. Int J Adv Manuf Technol 76(5–8):1173–1184

Zhang Q, Luo J, Duan Y (2014) Buyer–vendor coordination for fixed lifetime product with quantity discount under finite production rate. Int J Syst Sci 47(4):821–834. https://doi.org/10.1080/00207721.2014.906684

# Chapter 9
# An Integrated and Collaborated Supply Chain Model Using Quantity Discount Policy with Back Order for Time Dependent Deteriorating Items

**Isha Talati, Poonam Mishra, and Azharuddin Shaikh**

**Abstract** Nowadays entrepreneurs need to examine the different ways to grow and survive in this competitive environment. Coordination among the players and the use of appropriate promotional tools can reduce the total cost of supply cost. In this we have studied the effect of quantity discount on independent as well as integrated scenario. Here we have considered time-dependent demand for time-dependent deteriorating items with a fixed lifetime under shortages. A numerical example is presented to support this inventory model. The proposed model is useful to the decision-making of the supply chain associated with Drugs, Cosmetics, FMCGs, etc.

**Keywords** Integrated inventory model · Time dependent demand · Time dependent deterioration · Fix life time product · Shortages · Quantity discount · Inspection policy

**MSC** 90B85 · 90C26

## 9.1 Introduction

For any business, it is very difficult to grow and survive in this competitive world. In this scenario co-ordination among business players can reduce the total cost of supply chain and hence can increase over all supply chain profit. Goyal (1976) firstly developed integrated model for supplier-customer. Banerjee (1986) optimized integrated

I. Talati (✉)
Department of Engineering and Physical Sciences, Institute of Advanced Research, Koba Institutional Area, Gandhinagar 382426, India

P. Mishra
Department of Mathematics, School of Technology, Pandit Deendayal Petroleum University, Raisan Gandhinagar 382007, India

A. Shaikh
Institute of Management, Nirma University, Sarkhej-Gandhinagar Highway, Ahmedabad 382481, India

 
N. H. Shah et al. (eds.), *Decision Making in Inventory Management*, Inventory Optimization, https://doi.org/10.1007/978-981-16-1729-4_9

inventory supply chain model for purchaser and vendor. Goyal and Gunasekaran (1995) and Rau et al. (2006) generalized this model for deteriorating items and three-echelon supply chain. Chung and Cárdenas-Barrón (2013) formulated integrated inventory model for deteriorating items under two-level trade credit. Further, Shah et al. (2015) optimized selling price and order quantity with price sensitive and trended demand. Talati and Mishra (2019) optimized cycle time and quantity using analytical and genetic approach.

In classical EOQ model constant demand rate was taken implicitly. In realistic situation demand rate is always not a constant function. It is varying with respect to time. The fundamental result in EOQ with time varying demand pattern was derived by Donaldson (1977). Dave and Patel (1981) generalized this model for deteriorating items. Wee and Wang (1999) considered time varying demand and formulated a variable production policy. Mishra and Singh (2011) minimized inventory cost in which the demand and holding cost is depending on time under partial backlogging. Further Mishra et al. (2013) generalized this model for time varying deterioration. To enjoy better profit on large quantities, manufacturer offers quantity discount to that retailer which accelerates overall demand and gives a win-win situation to both parties. Firstly, Monahan (1984) had increased vendor's profit by incorporating quantity discount policy. Further, this model extended by Chang et al. (2010) for the deteriorating items. Duan et al. (2010) used quantity discount and proved theoretically that it reduced the total inventory cost.This model was extended with more realistic assumptions by different researchers like Ravithammal et al. (2014), Ravithammal et al. (2014b), Pal and Chandra (2014), Sarkar (2016), Mishra and Talati (2018) etc.

In classical model, inventory depletes due to demand only. But in the real world, inventory of deteriorating units depletes with demand as well as deterioration. Ghare and Schrader (1963) pioneer derived EOQ model for exponential deteriorating items. Philip and Covert (1973) formulated model for time dependent deteriorating items. This model further generalized by Philip (1974). In real world, some of the items have their expiry date. Thus, items deterioration rate is proposal to time and maximum life of product. Chen and Teng (2014) took maximum life of product in account and optimized order quantity of retailer under trade credit. Wu et al. (2014) extended this model for two level trade credits. Teng et al. (2016) formulated model for time varying deteriorating items which has a fix life time and optimized order quantity under advanced payment. In this model, Wu et al. (2017) examined effect of advance-cash-credit scheme. Shah (2017) explained EOQ model for fix life time products and optimized order profit, selling price, cycle time and quantity under order linked trade credit. Mishra and Talati (2017) analysed problem for time dependent deteriorating items with fix life time and derived integrated optimal policy under random input. Jani et al. (2020) formulated inventory control policies for item that deteriorate with respect to time under two level order linked trade credit. Shah and Naik (2020) derived inventory model for deteriorating items with time-price backlog dependent demand.

## 9.2 Notations and Assumptions

### 9.2.1 Notations

#### 9.2.1.1 Inventory Parameters Related to Manufacturer

| | |
|---|---|
| $m_1$ | Order multiple in Model-1 |
| $m_2$ | Order multiple in Model-2 |
| $C_{cd}$ | Inspection cost per delivery (Variable) |
| $C_{if}$ | Inspection cost $/product lot (Fix) |
| $D$ | Time dependent demand |
| $A_m$ | Set up costs ($) |
| $a$ | Constant part of demand |
| $C_{iu}$ | Inspection cost $/unit time inspected (/unit) |
| $b$ | Variable part of demand which vary with time |
| $P$ | Production rate |
| $\rho$ | Capacity utilization |
| $h_m$ | Holding cost/unit/annum |
| $Q_m(t)$ | Economic order quantity per cycle |
| $L$ | The maximum life time of a product (in year) |
| $\text{TC}_m$ | Total cost for Model-1 |
| $\text{TC}_{wm}$ | Total cost in a Model-2 |

#### 9.2.1.2 Inventory Parameters Related to Retailer

| | |
|---|---|
| $n$ | Order multiple Model-1 |
| $\lambda$ | Order multiple in Model-2 and $\lambda Q_r(t)$ as the new quantity |
| $\pi$ | Back order cost |
| $A_r$ | Ordering costs ($) |
| $h_r$ | Holding cost/unit/annum |
| $B(\lambda)$ | Discount given by manufacturer if the retailer agree with change their order |
| $k$ | Back order rate for Model-1 |
| $k'$ | Back order rate for Model-2 |
| $Q_r(t)$ | Economic order quantity per cycle |
| $\text{TC}_r$ | Retailer's total cost in a Model-1 |
| $\text{TC}_{qr}$ | Retailer's total cost in a Model-2 |
| TC | System total cost in a Model-1 |
| $\text{TC}_q$ | System total cost in a Model-2 |

### *9.2.2 Assumptions*

1. In this model we have considered the supply chain of single manufacturer and single retailer.
2. Quantity discount being offered by manufacturer to the retailer if retailer agrees to change his order quantity by the fixed order quantity.
3. Shortages are allowed.
4. The back order rate is considered as a decision variable for the retailer.
5. The demand rate is linear function of time $D = a + bt : a, b > 0$.
6. The fix life products are deteriorating with respect to time and are defined as $\theta(t) = \frac{1}{1+L-t}$.
7. Three-level inspections take place to check the defectiveness of products at the manufacturer's end.
8. Production rate is constant.
9. We have considered instantaneous replenishment and negligible lead time.

## 9.3 Model Formulation

In this section, we have considered manufacturer produce items in one-set-up but ships after a fixed time through multiple deliveries (single-set-up-multiple delivery (SSMD)). The shortages are considered for the retailer with the back order rate k. As per the scenario, we have formulated the following two models:

**Model-1**: Without quantity discount: No quantity discount being offered by manufacturer to retailer.

**Model-2**: With quantity discount: Quantity discount being offered by manufacturer to retailer as he agrees to change the order as per the manufacturer production.

### *9.3.1 Model-1: Without Quantity Discount*

#### 9.3.1.1 Manufacturer's Total Cost

Here we have considered constant production rate. So the manufacturer on hand inventory at any instantaneous time $t$ is shown in Fig. 9.1 and it's defined by the following differential equation

$$\frac{\mathrm{d}Q_m(t)}{\mathrm{d}t} + \frac{1}{1+L-t} Q_m(t) = P; 0 \leq t \leq T \tag{9.1}$$

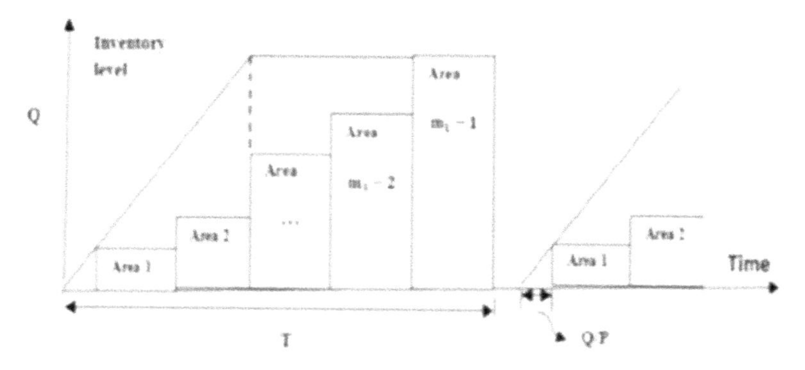


**Fig. 9.1** Inventory position for manufacturer

Using boundary condition $Q_m(T) = 0$ we get the solution of differential equation (9.1)

$$Q_m(t) = -P \ln|\frac{1 + L - T}{1 + L - t}|(1 + L - t) \tag{9.2}$$

The initial quantity at $t = 0$ is

$$Q_m(0) = -P \ln|\frac{1 + L - T}{1 + L}|(1 + L) \tag{9.3}$$

The basic costs are
(a) Setup cost: Here we have considered SSMD model and we have taken constant set up cost so we get

$$\mathrm{SC}_m = A_m \tag{9.4}$$

(b) Holding cost: Here we have calculated the manufacturer's inventory per cycle by taking difference of accumulated level of retailer and manufacturer. Thus, the manufacturer's holding cost for average inventory per unit time is given by

$$\mathrm{HC}_m = h_m \left[(m_1 - 1)(1 + \frac{P}{D}) + \frac{P}{D}\right] \int_0^T Q_m(t)\mathrm{d}t \tag{9.5}$$

Solving Eq. (9.5) we get

$$\mathrm{HC}_m = -P h_m a_{11}[(m_1 - 1)(1 + \rho) + \rho] \tag{9.6}$$

where

$$a_{11} = \ln|1 + L - T|\left(T + \mathrm{TL} - \frac{T^2}{2}\right) - \frac{(1+L)^2}{2}\left(\ln|1+L| - \frac{1}{2}\right) + \frac{(1+L-T)^2}{2}\left(\ln|1+L-T| - \frac{1}{2}\right)$$

(c) Inspection cost: Inspection cost consists of three various costs: fix inspection cost/lot, variable inspection cost/delivery and fix inspection cost/unit. Hence we get

$$\mathrm{IC}_m = -\frac{a + bt}{\mathrm{Pa}_{11}}\left(C_{cd} - Pa_{11}C_{imu} + \frac{1}{m_1}C_{if}\right) \tag{9.7}$$

where

$$a_{11} = \ln|1 + L - T|\left(T + TL - \frac{T^2}{2}\right) - \frac{(1+L)^2}{2}\left(\ln|1+L| - \frac{1}{2}\right) + \frac{(1+L-T)^2}{2}\left(\ln|1+L-T| - \frac{1}{2}\right)$$

Hence, manufacturer total cost

$$\begin{aligned}\mathrm{TC}_m(m_1) &= \frac{1}{T}(\mathrm{SC}_m + \mathrm{HC}_m + \mathrm{IC}_(m)) \\ &= \frac{1}{T}(A_m - Ph_m a_{11}[(m_1 - 1)(1+\rho) + \rho] \\ &\quad - \frac{a+bt}{Pa_{11}}\left(C_{cd} - Pa_{11}C_{imu} + \frac{1}{m_1}C_{if}\right)\end{aligned}$$

$$= \frac{1}{T}\left\{-\frac{(a+bt)C_{ivf}}{Pa_{11}m_1} - Ph_m a_{11}m_1(1+\rho) + Ph_m a_{11} - \frac{(a+bt)C_{cd}}{Pa_{11}} + (a+bt)C_{ivv} + A_m\right\} \tag{9.8}$$

$$\mathrm{TC}_m(m_1) = \frac{a_{21}}{m_1} + a_{22}m_1 + a_{23} \tag{9.9}$$

where

$$a_{21} = -\frac{(a+bt)C_{imu}}{TPa_{11}};\quad a_{22} = -\frac{Ph_m a_{11}(1+\rho)}{T};$$

$$a_{23} = \frac{Ph_m a_{11}}{T} - \frac{(a+bt)C_{cd}}{TPa_{11}} + \frac{(a+bt)C_{if}}{T} + \frac{A_m}{T}$$

Therefore manufacturer's total cost can be written as

$$\text{Min TC}_m(m_1) = \frac{a_{21}}{m_1} + a_{22}m_1 + a_{23}$$

$$\text{Subject to } m_1 t \leq L \tag{9.10}$$

$$m_1 \geq 1$$

The optimum value of $m_1$ is given by

$$m_1 = \sqrt{\frac{a_{21}}{a_{22}}} = \sqrt{\frac{(a+bt)C_{if}}{P^2 h_m a_{11}^2 (1+\rho)}} \tag{9.11}$$

If $m_1^*$ is the optimum value of $m_1$, then we get

$$m_1^* = \max\left\{\sqrt{\frac{(a+bt)C_{if}}{P^2 h_m a_{11}^2 (1+\rho)}}, 1\right\}$$

$$m_1^* \geq 1 \tag{9.12}$$

So the optimal cost for manufacturer with respect to $m_1$ is

$$\text{TC}_m(m_1) = 2\sqrt{a_{21}a_{22}} + a_{23}$$

**Theorem** *If $L \geq \psi$ then*

$$m_1^* = max\left\{\sqrt{\frac{(a+bt)C_{if}}{P^2 h_m a_{11}^2 (1+\rho)}}, []\frac{L}{\psi}\right\} \geq 1$$

*where $\psi = \frac{Q_m}{(a+bt)}$; $m_1^*$ be the optimum of (9.8) and $[]x$ is the least integer greater than or equal tox. $L \geq \psi$ Is to ensure that $m_1^* \geq 1$.*

***Proof*** Since $\text{TC}_m(m_1)$ is strictly convex in $m_1$ we have

$$\frac{d^2\text{TC}_m}{dm_1^2} = \frac{2a_{21}}{Tm_1^3} \geq 0$$

As shown in (9.10) $m_1^*$ is an optimum of equation of (9.8). Substitute $t = \frac{Q_m}{a+bt}$ in the first constraints of Eq. (9.8), then the following inequality we get

$$m_1 \frac{Q_m}{a+bt} \leq L$$

Consider $m_{12}^* = \frac{L}{\frac{Qm}{a+bt}} \geq 1$ because $L \geq \psi$
Therefore optimum shipment = $m_1^*$ if $m_1^* \leq m_{12}^*$ =$m_{12}^*$; otherwise. Where $\mathrm{TC}_m(m_1)$is a convex function. ∴. If $L \geq \psi$ then

$$m_1^* = \max\left\{\sqrt{\frac{(a+bt)C_{if}}{P^2 h_m a_{11}^2(1+\rho)}}, []\frac{L}{\psi}\right\} \geq 1$$

□

#### 9.3.1.2 Retailer's Total Cost

The on hand inventory of retailer is deplete with time dependent demand $D$ and deterioration rate $\theta(t)$ Hence, as shown in Fig. 9.2 the instantaneous state of inventory at any instant of time t is described by the following equation

$$\frac{dQ_r(t)}{dt} + \frac{1}{1+L-t}Q_r(t) = -D(t); 0 \leq t \leq 1-k \tag{9.13}$$

Using boundary condition $Q_r(1-k) = 0$ we get solution of differential equation (9.12)

$$Q_r(t) = \left[\{ln|\frac{1+L-t}{L+k}|(a+b(1+L))\} + b(t-1+k)\right][1+L-t] \tag{9.14}$$

The initial condition at t=0 is

$$Q_r = \left[\left\{ln|\frac{1+L}{L+k}|(a+b(1+L))\right\} + b(k-1)\right][1+L] \tag{9.15}$$

The basic costs are (a) Ordering cost: Here we consider constant ordering cost so

$$\mathrm{OC}_r = nA_r \tag{9.16}$$

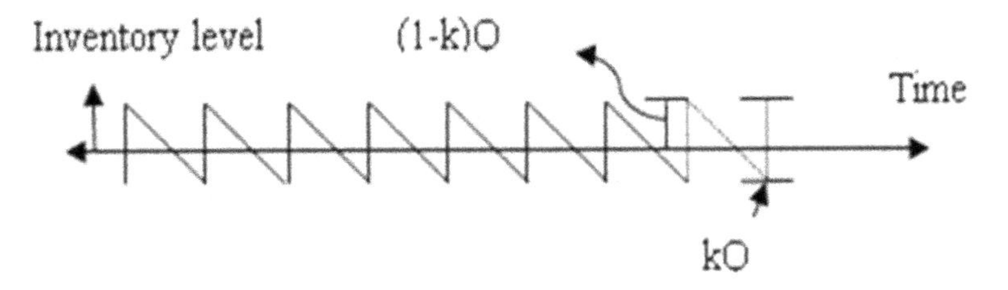


**Fig. 9.2** Inventory position for retailer

(b) Holding Cost: The holding cost for retailer's inventory level in the interval(0,1-k) is given by

$$\mathrm{HC}_r = h_r \int_0^1 Q_r(t)\mathrm{d}t \tag{9.17}$$

So,

$$\mathrm{HC}_r = h_r[(a + b(1 + L))(1 - k - (1 + L)ln|(L + k)(1 - L)|)] - \left[\frac{b}{2}(1 - k)^2\right]$$

(c) Back order Cost: The retailer's inventory level in the interval is given by

$$\mathrm{BC}_r = \pi \int_0^k Q_r(t)\mathrm{d}t \tag{9.18}$$

So,

$$\begin{aligned}\mathrm{BC}_r = \pi \{[(a + b(1 + L))((k - 1 - L) \ln|1 + L - k|) - k \ln|L + k| + (1 + L) \\ \ln|1 + L|] + \left[bk\left(\frac{3k}{2} - 1\right)\right]\}\end{aligned} \tag{9.19}$$

Thus retailer's total cost is

$$\mathrm{TC}_r(k) = \frac{1}{T}(\mathrm{OC}_r + \mathrm{HC}_r + \mathrm{BC}_r) \tag{9.20}$$

$$\text{Subject to } k > 0$$

So the integrated total cost

$$\mathrm{TC}(m_1, k) = \mathrm{TC}_m(m_1) + \mathrm{TC}_r(k) \tag{9.21}$$

## 9.3.2 *Model-2: With Quantity Discount*

In this model manufacturer requests the retailer to change his order size by a fix factor and offers retailer a quantity discount and retailer accept the offer. Thus the manufacturer and retailer new quantities are $\lambda m_2 Q_m$ and $\lambda a_r$ respectively.

### 9.3.2.1 Manufacturer's Total Cost

Here manufacturer requests the retailer to change his current order size by a factor $\lambda(> 0)$. Retailer accepts this offer and get quantity discount by discount factor B

($\lambda$). In this scenario the total cost for the manufacturer is

$$\mathrm{TC}_w m(m_2) = \frac{1}{T} 1/T(A_m - Ph_m a_1 1\lambda[(m_2 - 1)(1 + \rho) + \rho] \\ - \frac{(a + bt)}{P\lambda a_1 1}(C_i o - Pa_1 1 C_i mu + \frac{1}{m_2} C_i mp) + D(t)B())$$

Thus the optimization problem can be formulated as

$$\min \mathrm{TC}_q m(m_2) \tag{9.22}$$

$$\text{subject to } m_2 t \leq L, m_2 \geq 1$$

$$nA_r + \lambda h_r \left\{[(a + b(1 + L))(1 - k' - (1 + L)\ln|(L + k')(1 - L)|)] \right. \\ \left. - \left[\frac{b}{2}(1 - k')^2\right]\right\} + \lambda\pi \left\{[(a + b(1 + L))((k' - 1 - L)\ln|1 + L - k'|) - k\ln|L + k'| \right. \\ \left. +(1 + L)\ln|1 + L|] + \left[bk'(\frac{3k'}{2} - 1)\right]\right\} - TC_{wr}(m_2, k') \leq D(t)B(\lambda) \tag{9.23}$$

In Eq. (9.21), the first constraint of optimization problem represents that items are not overdue before they are used, and the third constraint term $DB(\lambda)$ represents compensation given by the manufacturer to the retailer to change the order.

#### 9.3.2.2 Retailer's Total Cost

According to the agreement between manufacturer and retailer, the retailer changes his order quantity. Thus the total cost for retailer with new quantity and quantity discount is

$$\mathrm{TC}_q r(k') = \frac{1}{T}\left\{nA_r + \lambda h_r \left\{[(a + b(1 + L))(1 - k' - (1 + L)\ln|(L + k')(1 - L)|)] \right.\right. \\ \left. -[\frac{b}{2}(1 - k')^2]\right\} + \lambda\pi \left\{[(a + b(1 + L))((k' - 1 - L)\ln|1 + L - k'|) \right. \\ \left. -k\ln|L + k'| + (1 + L)ln|1 + L|] + \left[bk'(\frac{3k'}{2} - 1)\right]\right\} \\ \left. -\mathrm{TC}_{wr}(m_2, k') \leq D(t)B(\lambda)\right\} \tag{9.24}$$

So the optimization problem can be formulated as

$$\min \mathrm{TC}_{qr}(k')$$

**Table 9.1** The optimal solution for without quantity discount optimization model

| Optimal | Independent scenario | Integrated scenario |
|---|---|---|
| Order size | 1.69 | 1.23 |
| Back order rate | 0.1835 | 0.1658 |
| Total cost | Independent scenario | Integrated scenario |
| Manufacturer's total cost | 10388.10 | 10220.21 |
| Retailer's total cost | 2070.16 | 1825.36 |
| Total supply chain cost | 12458.27 | 12045.57 |

$$\text{subject to } k' \geq 0$$

Thus the joint cost for quantity discount scenario is

$$\text{TC}_q(m_2, k') = \text{TC}_{qm}(m_2) + \text{TC}_{qr}(k')$$

## 9.4 Computational Algorithm

1. Set all parameters values in mathematical model.
2. Find optimal $m_1^*$ and $k^*$ from $\frac{\partial \text{TC}}{\partial m_1}$ and $\frac{\partial \text{TC}}{\partial k}$ simultaneously and obtain manufacturer and retailer total cost for integrated scenario. (Model-1: Without quantity discount)
3. Find optimal $m_2^*$ and $k^*$from $\frac{\partial \text{TC}_w}{\partial m_2}$ and $\frac{\partial \text{TC}_w}{\partial k'}$ simultaneously and obtain manufacturer and retailer total cost for joint scenario. (Model-2: With quantity discount).

## 9.5 Numerical Example

In this section numerical example is given. Consider $a = 400$, $b = 50$, $t = 0.5$ year, $= 0.1$, $C_{cd} = 1$\$/delivery, $C_{imu} = 0.02$\$/unit, $C_{if} = 20$\$/lot, $T = 0.7$ year, $h_m = 2$\$/unit, $L = 3$ years, $n = 5$, $A_r = 2000$\$, $A_m = 10000$\$, $P = 500$, $\pi = 5$\$, $h_r$ =2\$/unit
We have optimized this model using analytical method by MAPLE18. We get some computational results for Model 1 and Model 2 as shown in Tables 9.1 and 9.2 respectively.

In independent scenario of Model 1, the convexity of manufacturer's total cost and retailer's total cost are shown in Figs. 9.3 and 9.4 respectively.

**Table 9.2** The optimal solution for with quantity discount optimization model

| Optimal | Independent scenario | Integrated scenario |
|---|---|---|
| Order size | 5.37 | 5.23 |
| Back order rate | 0.1731 | 0.1584 |
| Total cost | Independent scenario | Integrated scenario |
| Manufacturer's total cost | 9441.66 | 8536.23 |
| Retailer's total cost | 1408.91 | 1345.25 |
| Total supply chain cost | 10850.57 | 9881.48 |

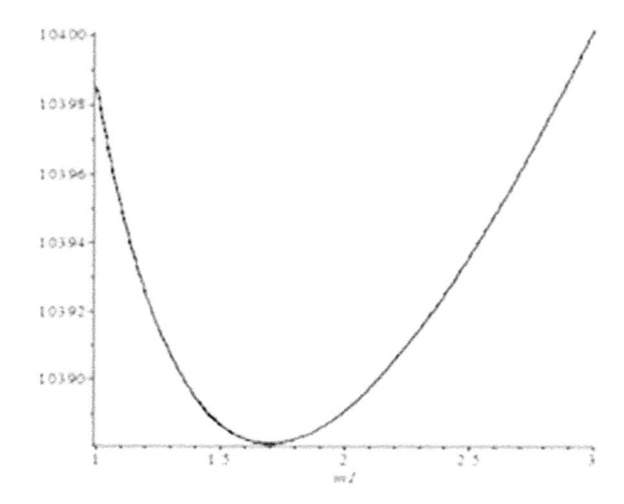


**Fig. 9.3** Convexity of manufacturer's cost for Model-1

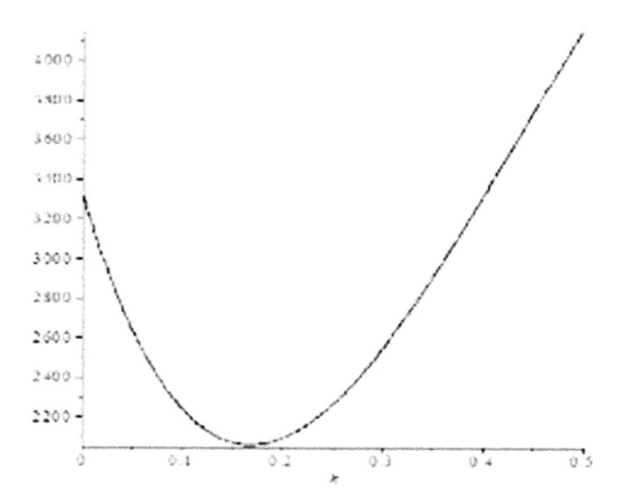


**Fig. 9.4** Convexity of retailer's cost for Model-1

In integrated scenario of Model 1, the convexity of integrated total cost is given below.

$$\begin{vmatrix} \frac{\partial^2 \text{TC}}{\partial k^2} & \frac{\partial^2 \text{TC}}{\partial k \partial m_1} \\ \frac{\partial^2 \text{TC}}{\partial k \partial m_1} & \frac{\partial^2 \text{TC}}{\partial m_1^2} \end{vmatrix} = 0.26352 \times 10^3$$

and $\frac{\partial^2 \text{TC}}{\partial k^2} = 0.4039621 \times 10^3 \geq 0$.

In independent scenario of Model 2, the convexity of manufacturer's total cost and retailer's total cost are shown in Figs. 9.5 and 9.6 respectively.

In integrated scenario of Model 2, the convexity of integrated total cost is given below.

$$\begin{vmatrix} \frac{\partial^2 \text{TC}}{\partial k'^2} & \frac{\partial^2 \text{TC}}{\partial k' \partial m_2} \\ \frac{\partial^2 \text{TC}}{\partial k' \partial m_2} & \frac{\partial^2 \text{TC}}{\partial m_2^2} \end{vmatrix} = 2.26352 \times 10^2 > 0$$

and $\frac{\partial^2 \text{TC}}{\partial k'^2} = 0.6039621 \times 10^3 \geq 0$.

**Observations**

- From Table 9.1 we have noticed that in the integrated scenario manufacturer total cost, retailer total cost and back order rate reduce as compare to independent scenario.
- It is clear from Table 9.2 that in the integrated scenario all costs and back order rate reduce as compare to independent scenario.
- By comparing Tables 9.1 and 9.2 we have observed in with co-ordination model manufacturer total cost, retailer total cost and back order rate reduce in both independent and integrated scenario.

## 9.6 Conclusion

This model is for items that deteriorating with respect to time but in fixed lifetime L under the time-dependent demand. The effect of quantity discount is demonstrated in the Model-2. From the numerical example, it is clear that the quantity discount policy reduces the back order rate, the total cost for the individual as well as the joint cost of the whole system. The convexity of the total cost function with respect to optimal order quantity and back order rate are studied graphically and mathematically in isolated as well as the integrated scenarios. Our result can help a retailer to decide to agree or disagree to change the order as per the manufacturer production because it reduces back order rate as well as the total cost. Hence, it increases in profit.

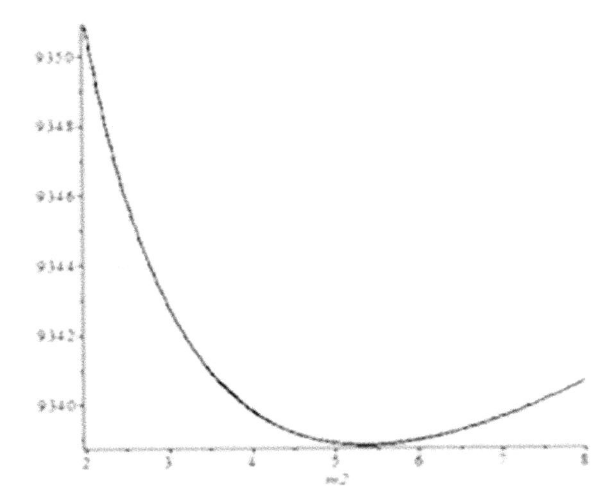


**Fig. 9.5** Convexity of retailer's cost for Model-1

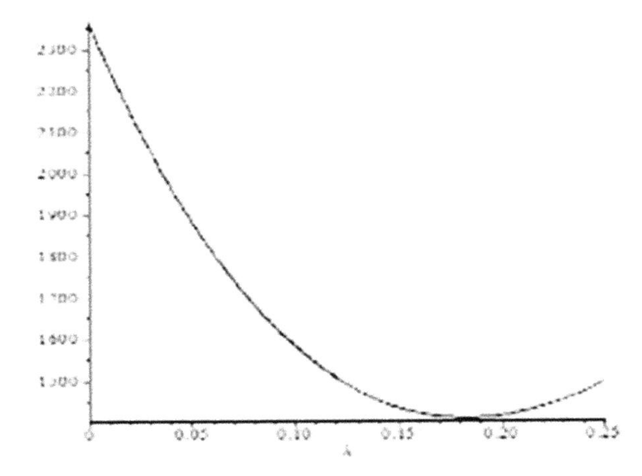


**Fig. 9.6** Convexity of retailer's cost for Model-1

## References

Banerjee A (1986) A joint economic-lot-size model for purchaser and vendor. Decis Sci 17(3):292–311

Chang C, Chen Y, Tsai R, Wu S (2010) Inventory models with stock-and price dependent demand for deteriorating items based on limited shelf space. Yugoslav J Oper Res 20(1):55–69

Chen SC, Teng JT (2014) Retailer's optimal ordering policy for deteriorating items with maximum lifetime under supplier's trade credit financing. Appl Math Model 38(15–16):4049–4061

Chung KJ, Cárdenas-Barrón LE (2013) The simplified solution procedure for deteriorating items under stock-dependent demand and two-level trade-credit in the supply chain management. Appl Math Model 37(7):4653–4660

Dave U, Patel LK (1981) (T, Si) policy inventory model for deteriorating items with time proposal demand. J Oper Res Soc 32:137–142

Donaldson WA (1977) Inventory replenishment policy for a linear trend in demand-an analytical solution. Oper Res Q 28(3):663–670

Duan Y, Luo J, Huo J (2010) Buyer-vendor inventory coordination with quantity discount incentive for fixed life time product. Int J Prod Econ 128:351–357

Ghare PM, Schrader GP (1963) A model for an exponentially decaying inventory. J Indu Eng 14(5):238–243

Goyal S (1976) An integrated inventory model for a single supplier-single customer problem. Int J Prod Res 15(1):107–111

Goyal S, Gunasekaran A (1995) An integrated production-inventory-marketing model for deteriorating items. Comput Ind Eng 28(4):755–762

Jani MY, Shah NH, Chaudhari U (2020) Inventory control policies for time-dependent deteriorating item with variable demand and two-level order linked trade credit. Optim Inventory Manage 55–67

Mishra VK, Singh LS, Kumar R (2013) An inventory model for deteriorating items with time dependent demand and time varying holding cost under partial backlogging. Int J Industr Eng 9(4):1–5

Mishra VK, Singh LS (2011) Deteriorating inventory model for time dependent demand and holding cost with partial backlogging. Int J Manage Sci Eng Manage 6(4):267–271

Mishra P, Talati I (2017) Integrated optimum policy with time dependent deterioration and random input for fix life time products. Math Today 33:88–98

Mishra P, Talati I (2018) Quantity discount for integrated supply chain model with backorder and controllable deterioration rate. Yugoslav J Oper Res 28(3):355–369

Monahan J (1984) A quantity discount pricing model to increase vendor profits. Manage Sci 30:720–726

Pal M, Chandra S (2014) A periodic review inventory model with stock dependent demand, permissible delay in payment and price discount on backorders. Yugoslav J Oper Res 24(1):99–110

Philip GC (1974) A generalized EOQ model for items with Weibull distribution. AIIE Trans 6:159–162

Philip GC, Covert RP (1973) An EOQ model for items with Weibull distribution deterioration. All E Trans 5(4):323–326

Rau H, Wu M, Wee H (2006) Integrated inventory model for deteriorating items under a multi-echelon supply chain environment. Int J Prod Econ 86:155–168

Ravithammal M, Uthayakumar R, Ganesh S (2014b) An integrated production inventory system for perishable items with fixed and linear back orders. Int J Math Anal 8(32):1549–1559

Ravithammal M, Uthayakumar R, Ganesh S (2014a) Buyer-vendor incentive inventory model with fix life time product with fixed and linear back order. Nat J Adv Comput Manage 5(1):21–34

Sarkar B (2016) Supply chain coordination with variable backorder, inspections, and discount policy for fixed lifetime product. Mathe Probl Eng 1–14

Shah NH (2017) Retailer's optimal policies for deteriorating items with a fixed lifetime under order linked conditional trade credit. Uncertain Supply Chain Manage 5(2):126–134

Shah NH, Naik MK (2020) Inventory policies for deteriorating items with time-price backlog dependent demand. Int J Syst Sci: Oper Logist 7(1):76–89

Shah NH, Shah DB, Patel DG (2015) Optimal pricing and ordering policies for inventory system with two-level trade credits under price-sensitive trended demand. Int J Appl Comput Math 1(1):101–110

Talati I, Mishra P (2019) Optimal production integrated inventory model with quadratic demand for deteriorating items under inflation using genetic algorithm. Investigación Oper 40(3)

Teng JT, Cárdenas-Barrón LE, Chang HJ, Wu J, Hu Y (2016) Inventory lot-size policies for deteriorating items with expiration dates and advance payments. Appl Math Model 40(19):8605–8616
Wee HM, Wang WT (1999) A variable production scheduling policy for deteriorating items with time varying demand. Comput Oper Res 26(3):237–254
Wu J, Ouyang LY, Cárdenas-Barrón LE, Goyal SK (2014) Optimal credit period and lot size for deteriorating items with expiration dates under two-level trade credit financing. Eur J Oper Res 237(3):898–908
Wu J, Teng JT, Chan YL (2017) Inventory policies for perishable products with expiration dates and advance-cash-credit payment schemes. Int J Syst Sci: Oper Logist 1–17

# Chapter 10
# A Wine Industry Inventory Model for Deteriorating Items with Two-Warehouse Under LOFO Dispatching Policy Using Particle Swarm Optimization

**Ajay Singh Yadav, Neha Chauhan, Navin Ahlawat, and Anupam Swami**

**Abstract** This article develops a deterministic inventory model for the wine industry for item spoilage with two storage systems and a time-dependent demand with partial bottlenecks. The inventory is transferred to OW according to the RW Bulk Discharge model and transport costs are considered negligible. The decline in the two camps is constant, but different due to the different storage methods. Use of particle swarm optimization as part of the LOFO shipping policy. The cost of ownership is considered constant up to a certain point in time and increases. Optimization of the particle swarm with different population sizes is second-hand to get to the bottom of the model. In this particle swarm optimization, a compartment of the best children is incorporated in the parent population for the next generation, and the size of that subset is a percentage of the size of their parent set. The digital sample is presented to make obvious the development of Modusland for corroboration. The kindliness analysis is performed separately for each parameter.

**Keywords** Two-warehouses · Instantaneous deterioration · Variable holding cost · Last in First out (LOFO) method · Particle swarm optimization

---

A. S. Yadav
Department of Mathematics, SRM Institute of Science and Technology, Delhi-NCR Campus, Ghaziabad, U.P., India

N. Chauhan (✉)
SRM Institute of Science and Technology, Delhi-NCR Campus, Ghaziabad, U.P., India

N. Ahlawat
Department of Computer Science, SRM Institute of Science and Technology, Delhi-NCR Campus, Ghaziabad, U.P., India

A. Swami
Department of Mathematics, Government Post Graduate College, Sambhal, U.P., India



N. H. Shah et al. (eds.), *Decision Making in Inventory Management*, Inventory Optimization, https://doi.org/10.1007/978-981-16-1729-4_10

## 10.1 Introduction

The classic models of wine industry inventories are mainly urbanized with the unique storage structure. In the earlier period, researchers have done a lot of investigation on inventory management in the wine industry and inventory management organization for the wine industry. The wine industry's inventory management and control system focuses primarily on issues and issues in the supply chain. To this end, production units (manufacturers of end products), sellers, supplier, and retailer require raw materials, finished products for prospect supply and demand on the market and at the customer. conventional models assume that demand and maintenance costs are invariable and that goods are delivered immediately when needed under an infinite supply policy. However, many scientists have suggested that over time, demand may change over time due to price and other factors, and that storage costs may change over time and other factors. Many models have been developed with different time-dependent needs in mind, with and without failures. All models that take into account fluctuations in demand in response to stock levels in the wine industry assume that the cost of ownership is constant throughout the cycle of stocks in the wine industry. Studies of stock models in the wine industry often assume unlimited storage capacity. In crowded markets such as supermarkets, corporate markets, etc., storage space for items may be limited. Another case of insufficient storage space can occur when you decide to buy a large number of items. This may be due to an attractive wholesale discount, or when the cost of purchasing goods exceeds the cost of inventory in another wine industry, or when the demand for goods is very high or very high for the goods in question. seasonal production is high, such as crop yields or problems with regular supplies. In this case, these items cannot be stored in an existing warehouse (your warehouse, OW for short). Therefore, in order to store excess items, you must find another warehouse (rented warehouse, abbreviated RW), which may be a short distance from EW or a little out of the way due to the unavailability of a nearby warehouse. is rented on a rent basis.

Particle swarm optimization is initiated from a population of random solutions and a random measure is assigned to each possible solution. Possible solutions, called particles, are blown through the problem area. Each particle follows its coordinates in the problem area, which indicate the best solution or suitability to date. The value of fitness is also recorded. This value is called pbest. The next best value recorded by the global version of PCA is the best total value and position of particles in the population so far. This value is called the best. Therefore, the particle changes velocity with each step and moves in the direction of Pbest and Gbest. This is a global version of PSO, where in addition to Pbest, each particle obtains the best solution called Nbest or Lbest from the local topological environment of the particles, this process is known as the local version of PSO.

## 10.2 Associated Works

Demand was believed to be a fluctuating function of time and that the backlog of unmet demand was a decreasing function of waiting time. Yadav and Swami (2018a, 2019a; Yadav et al. 2020d) "A model with a partial backlog in production inventory and lot size with time-varying operating costs and female decline". "Integrated supply chain model for material spoilage with linear demand based on inventory in an inaccurate and inflationary environment". "A flexible volume two-stage model with fluctuating demand and inflationary holding costs". Yadav et al. (2016, 2017a, 2019b, 2020a, e; Yadav and Swami 2018b) "Supply chain inventory model for two warehouses with soft IT optimization". "Multi-objective optimization for the stock model of electronic components and the degradation of double-bearing elements using a genetic algorithm". "An inflation inventory model for spoilage under two storage systems". "Chemical industry supply chain for warehouses with distribution centers using the Artificial Bee Colony algorithm". Management of the supply chain for electronic components of industrial electronics development for warehouses and their environmental impact using the particle swarm optimization algorithm". "Cost method for reliability considerations for the LOFO inventory model with warehouse for chemical industry". Pandey et al. (Yadav et al. 2020b) "An analysis of the inventory optimization of the marble industry based on genetic algorithms and particle swarm optimization". Malik et al. (Yadav et al. 2020c) "Security mechanism implemented in gateway service providers". Yadav et al. (2020f, g, h, i, j) "proposed the supply chain management of the National Blood Bank Center for the application of blockchain using a genetic algorithm". "Provided drug industry supply chain management for blockchain applications using artificial neural networks". "Suggested the red wine industry to manage the supply chain of distribution centers using neural networks". "A supply chain management for the rosé wine industry for storage using a genetic algorithm. "Providing supply chain management for the white wine industry for warehouses using neural networks". Chauhan and Yadav (Yadav et al. 2017b, 2020k) "proposed a stock model for commodity spoilage where demand depends on two stocks and stocks using a genetic algorithm". "Provide a car inventory system for inflation based on demand and inventory with a two-way distribution center using a genetic algorithm". Yadav et al. (2017c, 2020l, m, n, o; Yadav and Swami 2019c) "A method for calculating the reliability of the LIFO stock model with bearings in the chemical industry". "A Ensuring the management of the supply chain of electronic components for the development of the electronics industry in warehouses and the impact on the environment using the particle swarm optimization algorithm". "FIFO in Electrical Component Industry Green Supply Chain Inventory Model with Distribution Centers Using Particle Swarm Optimization". "LIFO in Automotive Components Industry Green Supply Chain Inventory Model with Bearings using Differential Evolution". "FIFO & LIFO in the Industry Green Supply Chain Inventory Model for Hazardous Substance Components with Storage using Simulated Annealing". "Health inventory control systems for blood bank storage with reliability applications using a genetic algorithm". Sana (Sana 2015, 2020)

"Price competition between green and non-green products in the context of a socially responsible retail and consumer services business magazine". "An EOQ model for stochastic demand for limited storage capacity". Moghdani et al. (2020) "Fuzzy model for economic production quantity with multiple items and multiple deliveries". Haseli et al. (2020) "Basic criterion for the multi-criteria decision-making method and its applications". Ameri et al. (2019) "Self-assessment of parallel network systems with intuitionistic fuzzy data: a case study". Birjandi et al. (2019) "Assessment and selection of the contractor when submitting a tender with incomplete information according to the MCGDM method". Gholami et al. (2018) "ABC analysis of clients using axiomatic design and incomplete estimated meaning". Jamali et al. (2018) "Hybrid Improved Cuckoo Search Algorithm and Genetic Algorithm to Solve Marko Modulated Demand".

## 10.3 Assumption and Notations

"The mathematical model of the inventory model for two lager industries of article breakdown is based on the following scores and assumptions"

**Notations**

$C_A$ = "cost of ordering under LOFO dispatching policy".

$\aleph_{ow}$ = "The ability of OW under LOFO dispatching policy".

$\aleph_{rw}$ = "The ability of RW under LOFO dispatching policy".

$T_{Ln}$ = "The length of replenishment cycle under LOFO dispatching policy".

$Q_{\max}$ = "Maximum wine industry Inventory level per cycle to be ordered".

$T_{L1}$ = "The time up to which wine industry inventory vanishes in RW".

$T_{L2}$ = "The time at which wine industry inventory level reaches to zero in OW and shortages begins".

$\omega$ = "Definite time up to which holding cost is constant under LOFO dispatching policy".

$\mathbb{N}_{ow}$ = "The holding cost of time per unit in OW under LOFO dispatching policy".

$\mathbb{N}_{rw}$ = "The holding cost of time per unit in RW under LOFO dispatching policy".

$\eta_0$ = "Data collection from RW to OW under LOFO dispatching policy".

$\eta$ = "Software Maintenance from RW to OW under LOFO dispatching policy".

$\begin{pmatrix} \text{LIFO} \\ \Pi \\ \text{RW} \end{pmatrix}$ = "wine industry inventory in RW of the level".

$\begin{pmatrix} \text{LIFO} \\ \Pi \\ \text{OW}(i) \end{pmatrix}$ = "wine industry inventory in OW of the level where, i = 1, 2".

$\begin{pmatrix} \text{LIFO} \\ \Pi \\ S \end{pmatrix}$ = "wine industry inventory level at time t in which the product has shortages".

$(\lambda + \lambda_0)$ = "Cost of Deterioration in RW under LIFO dispatching policy".

$(\lambda_1 + \lambda_2)$ = "Cost of Deterioration in OW under LIFO dispatching policy".

$\mathbb{N}_{P_c}$ = "Cost of Purchase per unit of items under LIFO dispatching policy".

$\mathbb{N}_{L_c}$ = "Cost of opportunity of time per unit under LIFO dispatching policy".

$\mathbb{N}_{S_c}$ = "Cost of shortages of time per unit under LIFO dispatching policy".

IB = "Maximum amount of wine industry inventory backlogged under LIFO dispatching policy".

IL = "Amount of inventory lost under LIFO dispatching policy".

$D_c$(LIFO) = "Data collection cost given by~$A_1$ $\eta_0$ where $A_1 > 0$".

$SM_c$(LIFO) = "Software Maintenance cost given by $A_0$ $\eta$ where $A_0 > 0$".

$T_L^C[\{T_{L2}, T_{Ln}\}(\text{LIFO})]$ = "The total relevant wine industry inventory cost per unit time of inventory system under LIFO dispatching policy".

**Assumption**

1. D(t) = Demand rate where $D(t) = \begin{bmatrix} \delta_0 \text{ if } t = 0 \\ \delta_0 + (\delta_1 + \delta_2)t \text{ if } t > 0 \end{bmatrix}$
2. $\mathbb{N}_{rw}$ = "we have k a time moment before which holding cost is constant".

where $\mathbb{N}_{rw} = \begin{bmatrix} \mathbb{N}_{rw} \text{ if } t < \omega \\ \mathbb{N}_{rw} t \text{ if } t > \omega \end{bmatrix}$

## 10.4 Wine Industry Inventory Model Mathematical Formulation and Analysis

"During the time interval, $(0, T_{L1})$ the stock in RW decreases due to demand and deterioration and is governed by the following differential equation":

$$\frac{d\begin{pmatrix} \text{LIFO} \\ \Pi \\ \text{RW} \end{pmatrix}(t)}{dt} = \left[-\{\delta_0 + (\delta_1 + \delta_2)t\} - \left((\lambda + \lambda_0)\begin{pmatrix} \text{LIFO} \\ \Pi \\ \text{RW} \end{pmatrix}(t)\right)\right] \quad 0 < t < T_{L1} \tag{10.1}$$

"In the time interval (0, $T_{L2}$) the stock level decreases in EV decrease only due to deterioration and is determined by differential equation"

$$\frac{d\begin{pmatrix}\text{LIFO}\\ \Pi\\ \text{OW1}\end{pmatrix}(t)}{dt} = \left[-(\lambda_1+\lambda_2)\begin{pmatrix}\text{LIFO}\\ \Pi\\ \text{OW1}\end{pmatrix}(t)\right] \quad 0 < t < T_{L1} \tag{10.2}$$

"During the time interval $(T_{L1}, T_{L2})$ the stock level in EV decreases due to both demand and deterioration, and is determined by the following differential equation"

$$\frac{d\begin{pmatrix}\text{LIFO}\\ \Pi\\ \text{OW2}\end{pmatrix}(t)}{dt} = \left[-\{\delta_0 + (\delta_1+\delta_2)t\} - \left((\lambda_1+\lambda_2)\begin{pmatrix}\text{LIFO}\\ \Pi\\ \text{OW2}\end{pmatrix}(t)\right)\right] \quad T_{L1} < t < T_{L2} \tag{10.3}$$

"Now at $t = T_{L2}$ the stock level disappears and the shortages arise in the time interval $T_{L2}, T_{Ln}T_{L2} = T_{Ln}$ a fraction of the total shortages is withheld and the shortage quantity is delivered to the customers at the beginning of the next replenishment cycle. The deficits are determined by the differential equation"

$$\frac{d\begin{pmatrix}\text{LIFO}\\ \Pi\\ S\end{pmatrix}(t)}{dt} = [-f\{\delta_0 + (\delta_1+\delta_2)t\}] \quad T_{L3} < t < T_{Ln} \tag{10.4}$$

"At the time $t = T_{Ln}$ the replenishment cycle is restarted. The aim of the model is to keep the total inventory costs as low as possible through the relevant costs".

"Now the stock level is given at different time intervals by solving the above differential equations (10.1)–(10.4) under boundary conditions"

$$\begin{pmatrix}\text{LIFO}\\ \Pi\\ \text{RW}\end{pmatrix}(T_{L1}) = 0;\ \begin{pmatrix}\text{LIFO}\\ \Pi\\ \text{OW1}\end{pmatrix}(0) = \aleph_{\text{ow}}$$

$$\begin{pmatrix}\text{LIFO}\\ \Pi\\ \text{OW2}\end{pmatrix}(T_{L2}) = 0;\ \begin{pmatrix}\text{LIFO}\\ \Pi\\ S\end{pmatrix}(t_2) = 0$$

Therefore Differential Eq. (10.1) gives

$$\begin{pmatrix}\text{LIFO}\\ \Pi\\ \text{RW}\end{pmatrix}(t) = \left\{\begin{array}{l}\left[\frac{\delta_0}{(\lambda+\lambda_0)} + \frac{(\delta_1+\delta_2)}{(\lambda+\lambda_0)^2}((\lambda+\lambda_0)T_{L1} - 1)e^{(\lambda+\lambda_0)(T_{L1}-t)}\right]\\ -\left[\frac{\delta_0}{(\lambda+\lambda_0)} + \frac{(\delta_1+\delta_2)}{(\lambda+\lambda_0)^2}((\lambda+\lambda_0)t - 1)\right]\end{array}\right\} \tag{10.5}$$

$$\begin{pmatrix}\text{LIFO}\\ \Pi\\ \text{OW1}\end{pmatrix}(t) = \aleph_{\text{ow}}e^{-(\lambda_1+\lambda_2)T_{L1}} \tag{10.6}$$

$$\begin{pmatrix} \text{LIFO} \\ \Pi \\ \text{OW2} \end{pmatrix}(t) = \left\{ \begin{array}{l} \left[ \dfrac{\delta_0}{(\lambda_1 + \lambda_2)} + \dfrac{(\delta_1 + \delta_2)}{(\lambda_1 + \lambda_2)^2} \{(\lambda_1 + \lambda_2) T_{L2} - 1\} e^{(\lambda_1 + \lambda_2)(T_{L2} - t)} \right] \\ - \left[ \dfrac{\delta_0}{(\lambda_1 + \lambda_2)} + \dfrac{(\delta_1 + \delta_2)}{(\lambda_1 + \lambda_2)^2} ((\lambda_1 + \lambda_2) t - 1) \right] \end{array} \right\} \quad (10.7)$$

$$\begin{pmatrix} LIFO \\ \Pi \\ S \end{pmatrix}(t) = f\left[ \delta_0 (T_{L2} - t) + \frac{(\delta_1 + \delta_2)}{2} (T_{L2}^2 - t^2) \right] \quad (10.8)$$

Now at $t = 0, \begin{pmatrix} \text{LIFO} \\ \Pi \\ \text{RW} \end{pmatrix}(0) = \aleph_{\text{rw}}$ therefore Eq. (10.5) yield

$$\aleph_{\text{rw}} = \left[ \begin{array}{l} \left( \dfrac{(\delta_1 + \delta_2)}{(\lambda + \lambda_0)^2} - \dfrac{\delta_0}{(\lambda + \lambda_0)} \right) \\ + \left( \dfrac{\delta_0}{(\lambda + \lambda_0)} + \dfrac{(\delta_1 + \delta_2)}{(\lambda + \lambda_0)^2} ((\lambda + \lambda_0) T_{L1} - 1) \mathrm{e}^{-(\lambda + \lambda_0) T_{L1}} \right) \end{array} \right] \quad (10.9)$$

"Maximum amount of inventory backlog during shortage period ($t = T_{Ln}$) is given by"

$$\begin{aligned} \text{IB} &= -\begin{pmatrix} \text{LIFO} \\ \Pi \\ S \end{pmatrix}(T_{Ln}) \\ &= f\left[ \delta_0 (T_{Ln} - t) + \frac{(\delta_1 + \delta_2)}{2} (T_{Ln}^2 - t^2) \right] \end{aligned} \quad (10.10)$$

"Quantity of inventory lost during a period of shortages"

$$LI = (1 - IB) = \left[ 1 - f\left\{ \delta_0 (T_{Ln} - t) + \frac{(\delta_1 + \delta_2)}{2} (T_{Ln}^2 - t^2) \right\} \right] \quad (10.11)$$

"The maximum stock that can be ordered is given as"

$$= \left[ \begin{array}{l} \aleph_{\text{ow}} + \left\{ \begin{array}{l} \left( \dfrac{(\delta_1 + \delta_2)}{(\lambda + \lambda_0)^2} - \dfrac{\delta_0}{(\lambda + \lambda_0)} \right) \\ + \left[ \dfrac{\delta_0}{(\lambda + \lambda_0)} + \dfrac{(\delta_1 + \delta_2)}{(\lambda + \lambda_0)^2} \right] ((\lambda + \lambda_0) T_{L1} - 1) \mathrm{e}^{(\lambda + \lambda_0) T_{L1}} \end{array} \right\} \\ + f\left\{ \delta_0 (T_{Ln} - T_{L2}) + \dfrac{(\delta_1 + \delta_2)}{2} (T_{Ln} - T_{L2}^2) \right\} \end{array} \right] \quad (10.12)$$

"Now continuity at $t = (T_{L1})$ shows that $\begin{pmatrix} LIFO \\ \Pi \\ OW1 \end{pmatrix}(T_{L1}) = \begin{pmatrix} LIFO \\ \Pi \\ OW2 \end{pmatrix}(T_{L1})$ therefore from Eqs. (10.6) and (10.7) we have"

$$\begin{bmatrix} (\delta_1+\delta_2)(\lambda_1+\lambda_2)^2T_{L2}^2 - \delta_0(\lambda_1+\lambda_2)^2T_{L2} \\ -\big((\lambda_1+\lambda_2)^2(\aleph_{ow}+Z)+(\delta_1+\delta_2)-\delta_0(\lambda_1+\lambda_2)\big) \end{bmatrix} = 0 \tag{10.13}$$

where $Z = \left\{\frac{\delta_0}{(\lambda_1+\lambda_2)} + \frac{(\delta_1+\delta_2)}{(\lambda_1+\lambda_2)^2}((\lambda_1+\lambda_2)T_{L1}-1)\right\}e^{-(\lambda_1+\lambda_2)T_{L1}}$

"Which is quadratic in $(T_{L2})$ and further can be solved for $(T_{L2})$ in terms of $(T_{L1})$ i.e".

$$T_{L2} = \varphi(T_{L1}) \tag{10.14}$$

where $\varphi(T_{L1}) = \frac{-\delta_0^2(\lambda_1+\lambda_2)^4 \pm \sqrt{D}}{2(\delta_1+\delta_2)(\lambda_1+\lambda_2)^2}$

$$D = \begin{bmatrix} \delta_0^2(\lambda_1+\lambda_2)^4 \\ +4(\delta_1+\delta_2)(\lambda_1+\lambda_2)^2 \left\{ \begin{matrix} (\delta_1+\delta_2)-\delta_0(\lambda_1+\lambda_2) \\ +(\lambda_1+\lambda_2)^2 \begin{bmatrix} \aleph_{ow}+ \\ \left( \begin{matrix} \dfrac{\delta_0}{(\lambda_1+\lambda_2)} \\ +\dfrac{(\delta_1+\delta_2)}{(\lambda_1+\lambda_2)^2}\{(\lambda_1+\lambda_2)T_{L1}-1\} \end{matrix} \right) e^{-(\lambda_1+\lambda_2)T_{L1}} \end{bmatrix} \end{matrix} \right\} \end{bmatrix}$$

"Subsequently, the total relevant inventory cost per cycle includes the following parameters":

1. "Ordering cost under LOFO dispatching policy"

$$C_A(LIFO) = \mathbb{N}_{C_A}$$

2. The present worth holding cost $= H_C$

**Case-1: When $\omega < T_{Ln}$ and $0 \le \omega < T_{L1}$ in RW**

$$H_c(\text{LIFO}) = \begin{bmatrix} \int_0^{\omega} \mathbb{N}_{rw}\begin{pmatrix}\text{LIFO}\\ \Pi \\ \text{RW}\end{pmatrix}(t)dt + \int_k^{T_{L1}} \mathbb{N}_{rw}t\begin{pmatrix}\text{LIFO}\\ \Pi \\ \text{RW}\end{pmatrix}(t)dt \\ + \int_0^{T_{L1}} \mathbb{N}_{ow}\begin{pmatrix}\text{LIFO}\\ \Pi \\ \text{OW1}\end{pmatrix}(t)dt + \int_{T_{L1}}^{T_{L2}} \mathbb{N}_{ow}\begin{pmatrix}\text{LIFO}\\ \Pi \\ \text{OW2}\end{pmatrix}(t)dt \end{bmatrix}$$

**Case-2: When $\omega > T_{Ln}$**

$$H_c(\text{LIFO}) = \left[\int_0^{T_{L1}} \mathbb{N}_{rw}\begin{pmatrix}\text{LIFO}\\ \Pi \\ \text{RW}\end{pmatrix}(t)dt + \int_0^{T_{L1}} \mathbb{N}_{ow}\begin{pmatrix}\text{LIFO}\\ \Pi \\ \text{OW1}\end{pmatrix}(t)dt + \int_{t_{L1}}^{T_{L2}} \mathbb{N}_{ow}\begin{pmatrix}\text{LIFO}\\ \Pi \\ \text{OW1}\end{pmatrix}(t)dt\right]$$

**"Holding cost for Case -1"**

$$H_c(\text{LIFO}) = \left[\begin{array}{l} \mathbb{N}_{rw}\left(\begin{array}{l} \delta_0 T_{L1}\omega + (\delta_1+\delta_2)T_{L1}^2\omega - \dfrac{(\delta_1+\delta_2)\omega^2}{2(\lambda+\lambda_0)} \\ -\dfrac{(\delta_1+\delta_2)T_{L1}^2}{3(\lambda+\lambda_0)} + \delta_0 T_{L1}^3 + (\delta_1+\delta_2)T_{L1}^4 \\ -\dfrac{\delta_0\omega}{(\lambda+\lambda_0)} - (\delta_1+\delta_2)T_{L1}\omega^2 - \delta_0 T_{L1}\omega^2 \\ -(\delta_1+\delta_2)T_{L1}^2\omega^2 + \dfrac{(\delta_1+\delta_2)T_{L1}\omega^2}{(\lambda+\lambda_0)} \\ +\dfrac{\delta_0\omega^2}{(\lambda+\lambda_0)} + \dfrac{(\delta_1+\delta_2)\omega^3}{3(\lambda+\lambda_0)} \end{array}\right) \\ +\mathbb{N}_{ow}\left(\begin{array}{l} \aleph_{ow}T_{L1} + \dfrac{(\delta_1+\delta_2)T_{L2}^2}{(\lambda_1+\lambda_2)} - \dfrac{(\delta_1+\delta_2)T_{L1}T_{L2}}{(\lambda_1+\lambda_2)} \\ +\dfrac{(\delta_1+\delta_2)T_{L1}^2}{2(\lambda_1+\lambda_2)} - \dfrac{(\delta_1+\delta_2)T_{L2}^2}{2(\lambda_1+\lambda_2)} \end{array}\right) \end{array}\right] \quad (10.15)$$

**"Holding cost for Case -2"**

$$H_c(\text{LIFO}) = \left[\begin{array}{l} \mathbb{N}_{rw}\left(\begin{array}{l} \delta_0 T_{L1}^2 + (\delta_1+\delta_2)T_{L1}^3 \\ +\dfrac{(\delta_1+\delta_2)T_{L1}^2}{(\lambda+\lambda_0)} - \dfrac{(\delta_1+\delta_2)T_{L1}^2}{2(\lambda+\lambda_0)} \end{array}\right) \\ +\mathbb{N}_{rw}\left(\begin{array}{l} \aleph_{ow}T_{L1} + \dfrac{(\delta_1+\delta_2)T_{L2}^2}{(\lambda_1+\lambda_2)} - \dfrac{(\delta_1+\delta_2)T_{L1}T_{L2}}{(\lambda_1+\lambda_2)} \\ +\dfrac{(\delta_1+\delta_2)T_{L1}^2}{2(\lambda_1+\lambda_2)} - \dfrac{(\delta_1+\delta_2)T_{L2}^2}{2(\lambda_1+\lambda_2)} \end{array}\right) \end{array}\right] \quad (10.16)$$

3. "The present worth of shortages cost"

$$S_c(\text{LIFO}) = \mathbb{N}_{S_c} f \left(\begin{array}{l} \dfrac{\delta_0 T_{Ln}^2}{2} - \dfrac{\delta_0 T_{L2}^2}{2} + \dfrac{(\delta_1+\delta_2)T_{Ln}^3}{6} \\ -\dfrac{(\delta_1+\delta_2)T_{L2}^3}{6} - \delta_0 t_1 T + \delta_0 T_{L2}^2 \\ -\dfrac{(\delta_1+\delta_2)T_{L2}^2 T_{Ln}}{2} + \dfrac{(\delta_1+\delta_2)T_{L2}^3}{2} \end{array}\right) \quad (10.17)$$

4. "The present worth opportunity cost/Lost sale cost"

$$L_c(\text{LIFO}) = \mathbb{N}_{L_c}\left(1-\left(\begin{array}{l}\frac{\delta_0 T_{Ln}^2}{2}-\frac{\delta_0 T_{L2}^2}{2}+\frac{(\delta_1+\delta_2)T_{Ln}^3}{6}\\ -\frac{(\delta_1+\delta_2)T_{L2}^3}{6}-\delta_0 T_{L1}T_{Ln}+\delta_0 T_{L2}^2\\ -\frac{(\delta_1+\delta_2)T_{L2}^2 T_{Ln}}{2}+\frac{(\delta_1+\delta_2)T_{L2}^3}{2}\end{array}\right)\right) \tag{10.18}$$

5. "Present worth purchase cost"

$$P_c(\text{LIFO}) = \mathbb{N}_{P_c}\left(\begin{array}{l}\aleph_{\text{ow}}+\left(\frac{(\delta_1+\delta_2)}{(\lambda+\lambda_0)^2}-\frac{\delta_0}{(\lambda+\lambda_0)}\right)\\ +\left\{\frac{\delta_0}{(\lambda+\lambda_0)}+\frac{(\delta_1+\delta_2)}{(\lambda+\lambda_0)^2}(\delta_0 T_{L1}-1)\text{e}^{-\delta_0 T_{L1}}\right\}\\ +f\left\{\delta_0(T_{Ln}-T_{L2})+\frac{(\delta_1+\delta_2)}{2}\left(T_{Ln}^2-T_{L2}^2\right)\right\}\end{array}\right) \tag{10.19}$$

6. "The data collection cost is given asunder LOFO dispatching policy".

$$D_c(\text{LIFO}) = \text{A}_1\eta_0 \tag{10.20}$$

7. "The Software Maintenance cost is given asunder LOFO dispatching policy".

$$SM_c(\text{LIFO}) = \text{A}_0\eta \tag{10.21}$$

"Therefore total relevant wine industry inventory cost per unit per unit of time is denoted and given by.

**Case-1**

$$T_L^C[\{T_{L2},T_{Ln}\}(LIFO)] = \left[\begin{array}{l}C_A(LIFO)+H_c(LIFO)+S_c(LIFO)\\ +L_c(LIFO)+P_c(LIFO)+D_c(\text{LIFO})+SM_c(\text{LIFO})\end{array}\right] \tag{10.22}$$

**Case-2**

$$T_L^C[\{T_{L2},T_{Ln}\}(LIFO)] = \left[\begin{array}{l}C_A(LIFO)+H_c(LIFO)+S_c(LIFO)\\ +L_c(LIFO)+P_c(LIFO)+D_c(\text{LIFO})+SM_c(\text{LIFO})\end{array}\right] \tag{10.23}$$

## 10.5 Particle Swarm Optimization

1. "Partition the region which contains the optimal solution".
2. "Initialize all pheromone trails with the same amount of pheromone and randomly generate a feasible solution".
3. "Ant movement according to the pheromone trails to produce feasible solution".
4. "Repeat the third step for a given number of ants".
5. "Pheromone update according to the best feasible solution in the current algorithm iteration".
6. "Repeat the third to fifth steps for a given number of cycles or a terminate criterion".
7. "Report the best solution".

## 10.6 Numerical Analysis

The following data, randomly selected in the respective units, were used to find the optimal solution and verify the model of the three actors, the manufacturer, the distributor, and the retailer. The data are provided as $\delta_0 = 601$, $\mathbb{N}_{C_A} = 1601$, $\aleph_{rw} = 3001, \aleph_{ow} = 3011, \delta_1 = 7.01, \delta_2 = 5.01, \mathbb{N}_{ow} = 61, \mathbb{N}_{rw} = 71, \mathbb{N}_{P_c} = 1701$, $(\lambda + \lambda_0) = 0.013$, $(\lambda_1 + \lambda_2) = 0.104$, $\mathbb{N}_{S_c} = 261$, $\omega = 1.57, f = 0.05$,$A_0 = 1.26$, $\eta = 1.66$, $A_1 = 1.46$, $\eta_0 = 2.66$ and $\mathbb{N}_{L_c} = 301$. The values of the decision variables are calculated for the model separately for two cases. Computationally optimal model solutions are given in Table 10.1. Actual values must be adjusted to the specific ACOA through experience and trial and error. However, some default settings are mentioned in the literature. Population size = 160, number of generations = 1100, crossing type = colon, crossing rate = 1.7, mutation types = bit flip, mutation rate = 0.017 per bit, because point crossing is used instead of one transition point by two slices cross rate is can be reduced to a maximum of 1.60.

### *10.6.1 Numerical Comparison Between Two Cases of the Model*

Using the same value of the parameter from the numerical example, we obtain the total cost of the relevant inventory of the wine industry model for two cases, as shown in Table 10.1. The table shows that this model immediately deteriorates and partially lags in case 1. If the cost of ownership in RW vary over time, this is the most expensive condition. In case 2, where the cost of ownership does not vary according

**Table 10.1** Wine industry model for two cases

| Case | Cost function | $T_{L1}$ | $T_{L2}$ | $T_{Ln}$ | Total relevant cost | Particle swarm optimization |
|---|---|---|---|---|---|---|
| 1 | $T_L^C[\{T_{L2}, T_{Ln}\}(\text{LIFO})]$ | 6.47477 | 66.7057 | 74.6487 | 175,649 | 7.5649 |
| 2 | $T_L^C[\{T_{L2}, T_{Ln}\}(\text{LIFO})]$ | 7.71647 | 75.7460 | 77.9896 | 61,046.6 | 4.6568 |

to the length of the cycle, the model is the most flexible and meets the most favorable conditions. The mathematical software Geogebra is used.

## 10.7 Sensitivity Analysis

See Tables 10.2 and 10.3.

**Table 10.2** Sensitivity analysis in relation to all rates

| $\delta_0$ | $T_{L1}$ | $T_{L2}$ | $T_{Ln}$ | $T_L^C[\{T_{L2}, T_{Ln}\}(LIFO)]$ |
|---|---|---|---|---|
| 551 | 7.49618 | 74.1910 | 84.6198 | 640,800 |
| 601 | 7.51756 | 75.5790 | 85.0507 | 647,858 |
| 751 | 7.55006 | 79.7147 | 86.5811 | 651,940 |
| 451 | 7.44771 | 71.0971 | 87.4559 | 674,165 |
| 401 | 7.41651 | 69.7714 | 87.7677 | 670,564 |
| 651 | 7.1946 | 66.4114 | 84.7116 | 616,654 |
| $\delta_1 + \delta_2$ | $T_{L1}$ | $T_{L2}$ | $T_{Ln}$ | $T_L^C[\{T_{L2}, T_{Ln}\}(LIFO)]$ |
| 1.55 | 7.46488 | 69.4764 | 81.6867 | 647,476 |
| 1.60 | 7.45465 | 66.6486 | 79.4710 | 648,950 |
| 1.75 | 7.41861 | 59.8867 | 74.7741 | 667,695 |
| 1.45 | 7.48785 | 76.4417 | 87.6679 | 671,776 |
| 1.40 | 7.49601 | 80.8751 | 90.8861 | 664,674 |
| 1.65 | 7.50961 | 91.9646 | 99.1066 | 600,468 |
| $\mathbb{N}_{C_A}$ | $T_{L1}$ | $T_{L2}$ | $T_{Ln}$ | $T_L^C[\{T_{L2}, T_{Ln}\}(LIFO)]$ |
| 6651 | 7.47477 | 76.7066 | 84.6498 | 677,664 |
| 6801 | 7.47478 | 76.7071 | 84.6709 | 677,666 |
| 7651 | 7.47479 | 76.7098 | 84.6546 | 677,676 |
| 6751 | 7.47477 | 76.7045 | 84.6475 | 677,660 |

(continued)

**Table 10.2** (continued)

| $\delta_0$ | $T_{L1}$ | $T_{L2}$ | $T_{Ln}$ | $T_L^C[\{T_{L2}, T_{Ln}\}(LIFO)]$ |
|---|---|---|---|---|
| 6601 | 7.47475 | 76.7045 | 84.6475 | 677,660 |
| 851 | 7.47475 | 76.7009 | 84.6471 | 677,616 |
| $\mathbb{N}_{S_c}$ | $T_{L1}$ | $T_{L2}$ | $T_{Ln}$ | $T_L^C[\{T_{L2}, T_{Ln}\}(LIFO)]$ |
| 671 | 7.47605 | 57.0777 | 67.6067 | 678,686 |
| 701 | 7.47715 | 57.7166 | 67.0670 | 678,844 |
| 771 | 7.47967 | 57.9477 | 61.8675 | 640,167 |
| 661 | 7.47764 | 56.7148 | 65.0649 | 676,875 |
| 601 | 7.47178 | 51.8417 | 65.9866 | 675.887 |
| 161 | 7.46675 | 49.5406 | 70.9774 | 671,657 |
| $\mathbb{N}_{L_c}$ | $T_{L1}$ | $T_{L2}$ | $T_{Ln}$ | $T_L^C[\{T_{L2}, T_{Ln}\}(LIFO)]$ |
| 611 | 7.47468 | 56.6869 | 44.6606 | 677,576 |
| 661 | 7.47459 | 56.6604 | 44.6717 | 677,566 |
| 151 | 7.47477 | 56.5964 | 44.7056 | 677,771 |
| 96 | 7.47486 | 56.7678 | 44.6771 | 677,671 |
| 86 | 7.47494 | 56.7501 | 44.6654 | 677,761 |
| 56 | 7.47561 | 56.8169 | 44.1899 | 677,868 |
| $\omega$ | $T_{L1}$ | $T_{L2}$ | $T_{Ln}$ | $T_L^C[\{T_{L2}, T_{Ln}\}(LIFO)]$ |
| 1.771 | 7.56077 | 78.6187 | 91.7017 | 651,770 |
| 1.971 | 7.56961 | 84.4808 | 99.7418 | 666,740 |
| 6.411 | 7.77079 | 91.8567 | 616.8710 | 716,688 |
| 1.441 | 7.47617 | 67.1097 | 77.0706 | 664,166 |
| 1.681 | 7.79751 | 61.8680 | 70.6908 | 611,566 |
| 0.801 | 7.71061 | 49.5857 | 54.5768 | 76,565.8 |
| $\mathbb{N}_{P_c}$ | $T_{L1}$ | $T_{L2}$ | $T_{Ln}$ | $T_L^C[\{T_{L2}, T_{Ln}\}(LIFO)]$ |
| 1651 | 7.59686 | 57.9587 | 55.6051 | 640,618 |
| 1801 | 7.70547 | 55.1471 | 56.0761 | 646,667 |
| 6651 | 7.01070 | 58.7766 | 58.6698 | 649,574 |
| 1751 | 7.75049 | 51.7789 | 57.1999 | 674,868 |
| 1601 | 7.6189 | 59.9767 | 56.0506 | 671,945 |
| 751 | 6.76710 | 45.1799 | 57.8856 | 661,981 |
| $\mathbb{N}_{ow}$ | $T_{L1}$ | $T_{L2}$ | $T_{Ln}$ | $T_L^C[\{T_{L2}, T_{Ln}\}(LIFO)]$ |
| 61 | 7.46764 | 69.6488 | 81.9610 | 647,077 |
| 71 | 7.45096 | 66.9546 | 79.9066 | 648,606 |
| 91 | 7.41078 | 60.4577 | 75.6674 | 661,955 |
| 51 | 7.48546 | 76.6148 | 86.9845 | 671,789 |

(continued)

Table 10.2 (continued)

| $\delta_0$ | $T_{L1}$ | $T_{L2}$ | $T_{Ln}$ | $T_L^C[\{T_{L2}, T_{Ln}\}(LIFO)]$ |
|---|---|---|---|---|
| 41 | 7.49519 | 80.7077 | 90.6418 | 665,519 |
| 71 | 7.51657 | 98.4068 | 91.7874 | 607,176 |
| $\mathbb{N}_{rw}$ | $T_{L1}$ | $T_{L2}$ | $T_{Ln}$ | $T_L^C[\{T_{L2}, T_{Ln}\}(LIFO)]$ |
| 86.1 | 7.77176 | 74.7006 | 46.7956 | 646,706 |
| 90.1 | 7.68116 | 75.8104 | 48.4778 | 646,798 |
| 116.1 | 7.06749 | 79.9776 | 54.0750 | 659,476 |
| 67.1 | 7.5951 | 71.0109 | 41.9746 | 676,718 |
| 60.1 | 7.77747 | 69.1967 | 79.5458 | 667,578 |
| 77.1 | 4.79647 | 66.6479 | 70.8174 | 609,706 |
| $(\lambda + \lambda_0)$ | $T_{L1}$ | $T_{L2}$ | $T_{Ln}$ | $T_L^C[\{T_{L2}, T_{Ln}\}(LIFO)]$ |
| 0.0141 | 7.48566 | 71.0670 | 46.1171 | 677,746 |
| 0.0151 | 7.49444 | 69.5881 | 40.6887 | 670,418 |
| 0.0191 | 7.51704 | 66.6776 | 76.0787 | 666,777 |
| 0.0111 | 7.46111 | 74.6911 | 46.7690 | 646,616 |
| 0.0101 | 7.44745 | 77.0818 | 49.8010 | 647,756 |
| 0.0061 | 7.77759 | 88.6679 | 54.4048 | 674,561 |
| $(\lambda_1 + \lambda_2)$ | $T_{L1}$ | $T_{L2}$ | $T_{Ln}$ | $T_L^C[\{T_{L2}, T_{Ln}\}(LIFO)]$ |
| 0.0151 | 7.46516 | 86.0676 | 76.9065 | 676,687 |
| 0.0161 | 7.45647 | 89.1559 | 79.4509 | 168,687 |
| 0.0611 | 7.47479 | 97.6484 | 46.5066 | 617,490 |
| 0.0161 | 7.48558 | 69.1746 | 71.4646 | 647,601 |
| 0.0111 | 7.49781 | 65.7941 | 68.5791 | 649,575 |
| 0.001 | 7.54760 | 56.0107 | 18.7006 | 675,946 |

## 10.8 Conclusions

In this article, we have proposed a deterministic inventory model for the wine industry with two warehouses to break down items with linear demand as a function of time and with variable holding costs depending on the duration of the order cycle, in order to minimize the total inventory. cost of wine. as part of the shipping policy for LOFO particle swarm optimization. Bottlenecks are allowed and partially delayed. Two different cases have been discussed, one with variable holding costs throughout the cycle period and the other with constant holding costs throughout the cycle, and it is noted that during the variable holding costs the total inventory costs of the wine industry are much higher than in the other case. Use of particle swarm optimization as part of the LOFO shipping policy. Also, the proposed model is very useful for items that deteriorate a lot. Because as the deterioration rate in both warehouses increases, the overall inventory costs of the wine industry decrease. As part of the

**Table 10.3** Sensitivity analysis with PSO

| Function | Algorithm | Best | Worst | Mean | Standard deviation |
|---|---|---|---|---|---|
| $\delta_0$ | PSO | 0.10608 | 14.0900 | 64.6098 | 040,800 |
| $\delta_1 + \delta_2$ | PSO | 0.16488 | 19.4664 | 60.6806 | 043,436 |
| $\mathbb{N}_{C_A}$ | PSO | 0.16466 | 10.6060 | 64.0498 | 046,604 |
| $\mathbb{N}_{S_c}$ | PSO | 0.16468 | 60.6809 | 84.0600 | 036,160 |
| $\mathbb{N}_{L_c}$ | PSO | 0.16601 | 63.0333 | 83.6066 | 038,080 |
| $\omega$ | PSO | 0.10063 | 68.6083 | 80.6003 | 030,630 |
| $\mathbb{N}_{P_c}$ | PSO | 0.10608 | 14.0900 | 64.6098 | 040,800 |
| $\mathbb{N}_{ow}$ | PSO | 0.16488 | 19.4664 | 60.6806 | 043,436 |
| $\mathbb{N}_{rw}$ | PSO | 0.16466 | 10.6060 | 64.0498 | 046,604 |
| $(\lambda_1 + \lambda_2)$ | PSO | 0.16468 | 60.6809 | 84.0600 | 036,160 |
| $(\lambda + \lambda_0)$ | PSO | 0.16601 | 63.0333 | 83.6066 | 038,080 |

LOFO Shipping Directive, it uses Particle Swarm Optimization. This model can be further expanded to include other degradation rates, probabilistic demand models, and other realistic combinations using particle swarm optimization.

## References

Ameri Z, Sana SS, Sheikh R (2019) Self-assessment of parallel network systems with intuitionist fuzzy data: a case study. Soft Comput 23(23):12821–12832

Birjandi AK, Akhyani F, Sheikh R, Sana SS (2019) Evaluation and selecting the contractor in bidding with incomplete information using MCGDM method. Soft Comput 23(20):10569–10585

Gholami A, Sheikh R, Mizani N, Sana SS (2018) ABC analysis of the customers using axiomatic design and incomplete rough set. RAIRO-Oper Res 52(4–5):1219–1232

Haseli G, Sheikh R, Sana SS (2020) Base-criterion on multi-criteria decision-making method and its applications. Int J Manage Sci Eng Manage 15(2):79–88

Jamali G, Sana SS, Moghdani R (2018) Hybrid improved cuckoo search algorithm and genetic algorithm for solving Markov modulated demand. RAIRO-Oper Res 52(2):473–497

Moghdani R, Sana SS, Shahbandarzadeh H (2020) Multi-item fuzzy economic production quantity model with multiple deliveries. Soft Comput 24(14):10363–10387

Sana SS (2015) (2015) An EOQ model for stochastic demand for limited capacity of own warehouse. Ann Oper Res 233(1):383–399

Sana SS (2020) Price competition between green and non-green products under corporate social responsible firm. J Retail Consum Serv 55:102118

Yadav AS, Swami A (2018a) A partial backlogging production-inventory lot-size model with time-varying holding cost and weibull deterioration. Int J Procurement Manage 11(5)

Yadav AS, Swami A (2018b) Integrated supply chain model for deteriorating items with linear stock dependent demand under imprecise and inflationary environment. Int J Procurement Manage 11(6)

Yadav AS, Swami A (2019a) A volume flexible two-warehouse model with fluctuating demand and holding cost under inflation. Int J Procurement Manage 12(4):441–456

Yadav AS, Swami A (2019c) An inventory model for non-instantaneous deteriorating items with variable holding cost under two-storage. Int J Procurement Manage 12(6):690–710

Yadav AS, Mishra R, Kumar S, Yadav S (2016) Multi objective optimization for electronic component inventory model & deteriorating items with two-warehouse using genetic algorithm. Int J Control Theor Appl 9(2)

Yadav AS, Mahapatra RP, Sharma S, Swami A (2017a) An inflationary inventory model for deteriorating items under two storage systems. Int J Econ Res 14(9)

Yadav AS, Swami A, Kher G, Kumar S (2017b) Supply chain inventory model for two warehouses with soft computing optimization. Int J Appl Bus Econ Res 15(4):41–55

Yadav AS, Taygi B, Sharma S, Swami A (2017c) Effect of inflation on a two-warehouse inventory model for deteriorating items with time varying demand and shortages. Int J Procurement Manage 10(6)

Yadav AS, Kumar J, Malik M, Pandey T (2019b) Supply chain of chemical industry for warehouse with distribution centres using artificial bee colony algorithm. Int J Eng Adv Technol 8(2S2):14–19

Yadav AS, Pandey T, Ahlawat N, Agarwal S, Swami A (2020a) Rose wine industry of supply chain management for storage using genetic algorithm. Test Engraining Manage 83:11223–11230

Yadav AS, Ahlawat N, Agarwal S, Pandey T, Anupam Swami A (2020b) Red wine industry of supply chain management for distribution center using neural networks. Test Engraining Manage 83:11215–11222

Yadav AS, Ahlawat N, Dubey R, Pandey G, Swami A (2020c) Pulp and paper industry inventory control for Storage with wastewater treatment and Inorganic composition using genetic algorithm (ELD Problem). Test Engraining Manage 83:15508–15517

Yadav AS, Dubey R, Pandey G, Ahlawat N, Swami A (2020d) Distillery industry inventory control for storage with wastewater treatment & logistics using particle swarm optimization. Test Engraining Manage 83:15362–15370

Yadav AS, Navyata S, Ahlawat N, Swami A (2020e) Reliability consideration costing method for LOFO inventory model with chemical industry warehouse. Int J Adv Trends Comput Sci Eng 9(1):403–408

Yadav AS, Navyata S, Ahlawat N, Swami A (2020f) Reliability consideration costing method for LOFO Inventory model with chemical industry warehouse. Int J Adv Trends Comput Sci Eng 9(1):403–408

Yadav AS, Pandey P, Ahlawat N, Dubey R, Swami A (2020g) Wine industry inventory control for storage with wastewater treatment and pollution load using ant colony optimization algorithm. Test Engraining Manage 83:15528–15535

Yadav AS, Selva NS, Tandon A (2020h) Medicine manufacturing Industries supply chain management for Block chain application using artificial neural networks. Int J Adv Sci Technol 29(8s):1294–1301

Yadav AS, Swami A, Ahlawat N, Bhatt D, Kher G (2020i) Electronic components supply chain management of Electronic Industrial development for warehouse and its impact on the environment using Particle Swarm Optimization Algorithm. Int J Procurement Manage. In: Optimization and Inventory Management, pp 427–445

Yadav AS, Swami A, Ahlawat N, Bhatt D, Kher G (2020j) Electronic components supply chain management of electronic industrial development for warehouse and its impact on the environment using particle swarm optimization algorithm. Int J Procurement Manage. In: Optimization and inventory management, pp 427–445

Yadav AS, Tandon A, Selva NS (2020k) National blood bank centre supply chain management for blockchain application using genetic algorithm. Int J Adv Sci Technol 29(8s):1318–1324

Yadav AS, Bansal KK, Agarwal S, Vanaja R (2020l) FIFO in green supply chain inventory model of electrical components industry with distribution centres using particle swarm optimization. Adv Math Sci J 9(7):5115–5120

Yadav AS, Kumar A, Agarwal P, Kumar T, Vanaja R (2020m) LOFO in green supply chain inventory model of auto-components industry with warehouses using differential evolution. Adv Math Sci J 9(7):5121–5126

Yadav AS, Abid M, Bansal S, Tyagi SL, Kumar T (2020n) FIFO & LOFO in green supply chain inventory model of hazardous substance components industry with storage using simulated annealing. Adv Math Sci J 9(7):5127–5132

Yadav AS, Ahlawat N, Sharma N, Swami A, Navyata (2020o) Healthcare systems of inventory control for blood bank storage with reliability applications using genetic algorithm. Adv Math Sci J 9(7):5133–5142

# Chapter 11
# Integrated Lot Sizing Model for a Multi-type Container Return System with Shared Repair Facility and Possible Storage Constraint

**Olufemi Adetunji, Sarma V. S. Yadavalli, Rafid B. D. Al-Rikabi, and Makoena Sebatjane**

**Abstract** Containerisation has made global trade faster, more efficient, and safer, but it also comes with the challenge of container repositioning due to the imbalance between the container demand and supply in ports. This balancing act is necessary, not only at the aggregate level, but also the demands for the different types and sizes must be considered. In such system, we have containers that are returned after use, some of which are directly reused, some repaired and reissued, and some new ones purchased to make up for lost or badly damaged ones for each type and size of containers involved. There is, therefore, the need to have a model to plan the quantity and timing of the replenishment process for all the container varieties involved in an integrated manner. This paper develops a lot sizing model for a system in which these diverse types of containers are reused with the repair done in a common facility before they are reissued where necessary. A solution algorithm was devised by exploiting the structure of the problem, and a numerical example was provided to illustrate how to solve the problem using the algorithm. The solution is considered fit for ports management institutions that need to manage container return and repair systems in an integrated manner.

**Keywords** Lot sizing · Genetic algorithm · Container repositioning · Reparable inventory · Containerisation · Container return

## 11.1 Introduction

The advent of global supply chain management has changed the scale and scope of logistics. Driven by revolution in transportation, information technology, and change in the fiscal policies of many countries, many organizations have embraced global supply chain management, and ports have become central in making this a reality. Also, central to these port operations is the use of container as this makes handling

O. Adetunji (✉) · S. V. S. Yadavalli · R. B. D. Al-Rikabi · M. Sebatjane
Department of Industrial and Systems Engineering, University of Pretoria, Pretoria, South Africa
e-mail: olufemi.adetunji@up.ac.za

 
N. H. Shah et al. (eds.), *Decision Making in Inventory Management*,
Inventory Optimization, https://doi.org/10.1007/978-981-16-1729-4_11

of global trade items easier, safer, and more secure. An attendant problem that arises due to containerization is the management of empty containers that have been used to move goods across global and local destinations. There is usually the need to reposition many of these containers due to general imbalance in container demand and supply across most ports of the world.

A port hardly has a balanced container supply and demand. Most ports usually experience either surplus or shortage of containers depending on whether they are more import or export biased. This is common in most countries as some ports are closer to the industrial hubs and tend to ship out more items while some are closer to trading areas and bring in more containers. This often leads to situations where there is always the need to handshake between some surplus and other shortage ports within or between economies, which involves container movement and repositioning.

In this chapter, we consider a problem of container management in a port such that once containers are sent out, not all containers will return. A portion of the containers become lost or damaged and becomes unusable. Among the returned containers, a portion can be returned to service immediately (maybe after cleaning) while some other returned containers need to be fixed in the repair facility before they can be re-issued for use. The returned containers that are damaged but reparable are fixed and some more containers are bought from vendor to make up for the lost or permanently damaged containers. The cycle is then continuously repeated.

In such ports, there are many types of containers that need to be managed simultaneously. When balancing container demand and supply, it is important to not only meet the aggregate number of the required containers, but also the appropriate mix of each of the different types of containers (e.g. different sizes, and special requirements like reefers vs dry containers). The containers share repair facility and storage space. Management needs to decide how much space is allotted to each container type in the storage area. This is because storing containers in the port is not free and charges may be dependent on container types. There is the need to provide appropriate storage space for each container type and for all the containers. In addition, at the container repair facility, there may be equipment for repairing the containers that need to be set up depending on which type of containers is to be repaired. There is the need to schedule when to repair each container and how much of each type to repair in a cycle, so as to guarantee the varietal availability of containers. In addition, there is a need to decide when to buy top-up containers and how many to buy. Decision also needs to be made for repositioning of containers from surplus areas to areas of need. All these need to be jointly managed.

This cycle is shown in Fig. 11.1 where there is a demand at the rate, $D$, for containers per unit time. Of this demand, only a proportion, $x$, of the containers issued comes back for reuse. It is assumed $x$ is close to 1, and this is reasonable for most container management systems. When the containers return, a proportion of the returned containers needs to be repaired before they can be put back to use. This proportion is $y$ of the returned containers. Hence, the total number of returned containers is quantity $x.D$, of which $x.y.D$ go into repair while $(1-y).x.D$ is available to be reused immediately. There is also the need to buy some more containers, $(1-x).D$, to make up for the quantity that will not return for reuse in either serviceable

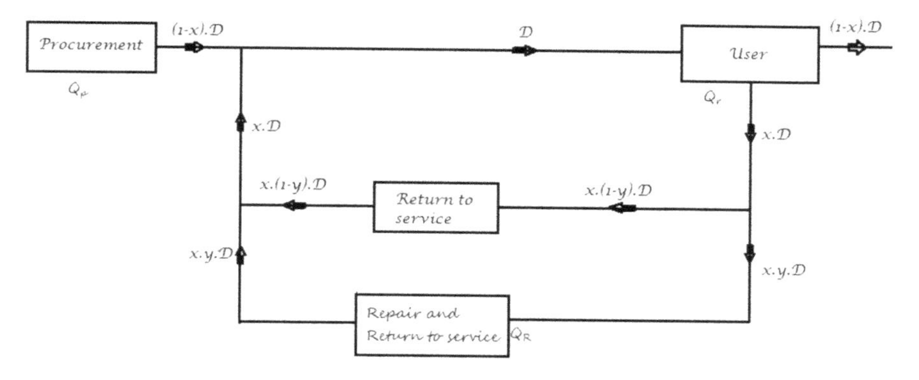


**Fig. 11.1** Container flow cycle with possible repair

or repairable form. A serviceable means the returned container can be immediately put to reuse without the need for any repair work, while the repairable return state means the returned container needs to first undergo some form of repair work before it can be put back into reuse. The lot size quantity for the containers procured each time is denoted by $Q_p$. When the containers are collected (repositioned), the lot size for each batch transferred is $Q_r$. When the repair is to be inducted, the lot size for a batch of repair is $Q_R$. The objective is to find the best lot size quantities, $Q_p$, $Q_r$ and $Q_R$ to minimize the total cost.

While container return systems can be classified as some form of the general return logistics problem, different reverse logistics systems exhibit some different behaviors, and such need to reflect in the modeling. Not much work has been seen in lot sizing of container return systems. Most reverse logistics models take their general form from those of repairable items, hence, inherited their assumptions as well. Schrady (1967) is a seminal work in this field. He distinguished between the continuous supplement policy and the substitution policy. While the replenishment triggers differ in the two cases, both however, assumed that the value of the 'Ready-For-Issue' (RFI) items is higher than that of 'Non-Ready-For-Issue' (NRFI) items, probably up to the order of five to one. The implication is that the RFI items are first utilized before the NRFI are considered in order to minimize the total holding cost. The thought here is, however, more like Koh et al. (2002) where consumption occurs simultaneously from both return and new items, and this is discussed next.

In the container repair problem presented, the management has to decide on the top-up procurement batch size, the return collection batch size, and the in-house repair batch size simultaneously. This presents a three-echelon problem as opposed to the two of the traditional repairable inventory system. In addition, the classic repairable inventory system assumes that items are returned continuously but drawn down during each repair batch. The draw down appears as steps in such models. In this problem, it is more realistic in the port's container management case that containers are considered to be returned in batches, and there is a continuous draw down for repair and reuse. This repair process has two phases: the phase where there is both repair and use of containers in which there is a gradual increase in inventory position (i.e. positive slope), followed by a use-only phase, which is a gradual decrease (negative slope). This is typical of all production-inventory systems. All these are shown in Fig. 11.2 for a single type of container. The topmost graph is that of top-up procurement with batch quantity, $Q_p$, being gradually consumed. There are some $m \geq 1$ integer cycles of return per single cycle of procurement. This is shown in the sub-graph "return cycles without repair" in Fig. 11.2. The returned containers are brought back in batch size $Q_r$. This graph looks so because the serviceable return containers are available for use immediately, and those that have just been repaired are also available for use immediately. All useable containers of a particular type are stored together where they may be issued, whether purchased, returned in usable form, or repaired and ready for re-use.

The sub-graph "repair cycles" in Fig. 11.2 shows the behavior of inventory of each inducted batch of repair. It should be noted that the height of this graph is $I_{\max,R}$ and not $Q_R$ because of simultaneous repair and use in the first phase of the inventory. There are $n \geq 1$ integer repair cycles for each return cycle. The sub-graph "return cycles with repair" in Fig. 11.2 maps the inventory position of the returned reparable return containers, a proportion of the returned quantity $Q_{,r}$ that is gradually drawn down for repair in batches of size $Q_R$, repeated $n$ times until it reaches zero. Hence, there are $n$ cycles of repair in a single cycle of return and $m \cdot n$ cycles of repair in a single cycle of procurement, so, if the cycle time of container repair is $T$, the cycle

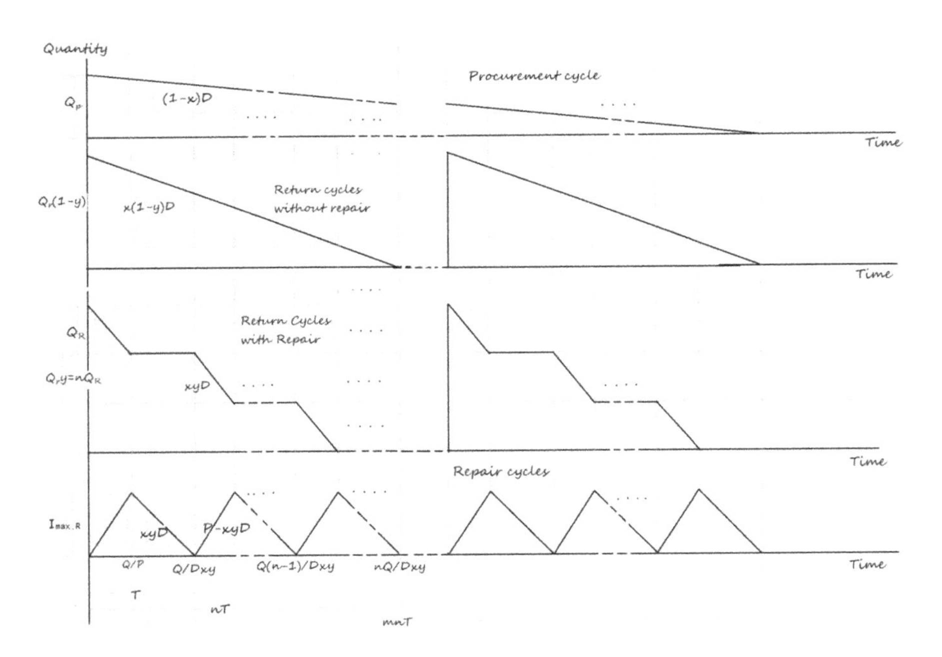


**Fig. 11.2** Inventory-time graphs for procurement, return and repair

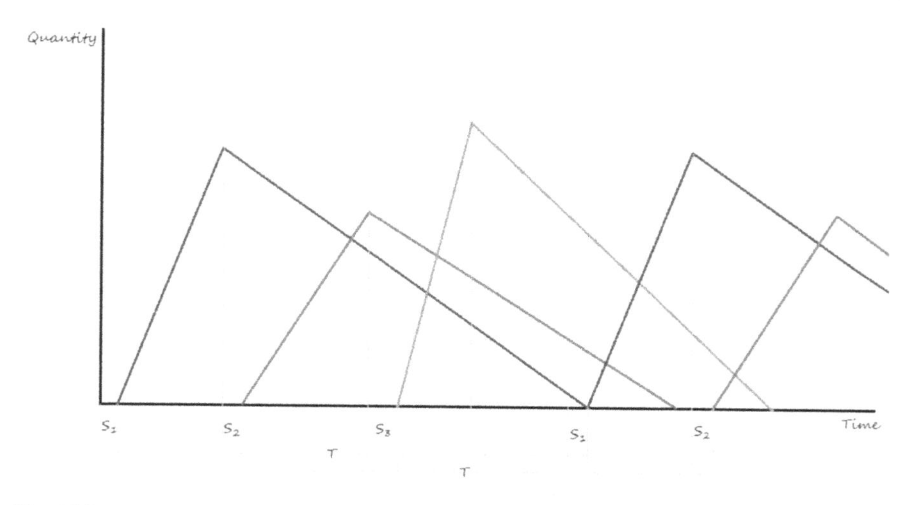


**Fig. 11.3** Integrated common repair cycle for a 3-container repair system

time for container collection (repositioning) will be $n \cdot T$, and that for procurement would be $m \cdot n \cdot T$. It can also be seen that also be seen that $Q_r \cdot y = nQ_R$.

While Fig. 11.2 shows the container stock position for a single type of container, the problem here considers more than one type of container, so, Fig. 11.2 would need to be overlaid for all types of containers involved. Since the cycle for repair integrates all types of containers because of shared repair and storage capacity, only the repair cycle diagram for a three-container system with a common repair centre is shown in Fig. 11.3 (without any loss of generalization). It indicates the common cycle time, $T$, that they all maintain in order to sequence the continuous repair cycle to accommodate each container type. This would be further discussed in the model development section. There is provision for a small setup (or changeover) time, $S$, for the repair of a batch of each type of container.

The sections in this paper are now summarised. The first section which is just concluded presents the introduction, where the problem background is provided, and the description of the problem is given. The next section is a brief review of the pertinent extant literature in order to identify the context of the solution provided. Following this is the section on model development where the integrated optimal lot sizing formula for this multi-echelon, multi-item return problem is developed. This is followed by a section presenting solution algorithm followed by an illustration of its use with numerical examples. The last section concludes, summarising the objectives and value of this work.

## 11.2 Literature Review

Containerization has had significant impact on global transportation and operations of ports, hence Rodrigue and Notteboom (2009) surmised it had the same effect on shipment as what the jet plane had on travel. Levinson (2016) said containerization has made the world smaller and economies bigger; moreover, it is a main catalyst of globalization. This method of movement of goods has flourished particularly in this dispensation of globalization as production of many items became regional while their consumption remains global (Yu et al. 2018). This regionalization and globalization made it necessary to find ways to move many items across borders efficiently and conveniently with little damage to products, and hence, containerization (Vodjani et al. 2013). Containerisation makes shipping more efficient and convenient as it greatly reduces handling and packing problems of materials shipped across borders. Most global goods movement has been via the use of container. A 2009 research by Lloyd's marine indicated that 75% by volume and 60% by value of global trade utilizes maritime transport for their movement, while 52% of maritime transport is by containers (Lee and Song 2016). This gives an indication of the volume of global goods moved via containers.

Containerisation has, however, come with its own challenges and the imbalance of shipment of containers is the main one. The demand for and supply of containers in all locations (ports and hinterland) are hardly equal, and this difference can be huge. Even at a global level, the difference in quantity of containers shipped and received between two destinations could be more than 100% of the other side as shown in Gencer (2019). This usually necessitates the need to reposition containers from places of surplus to places of deficit. Kuzmicz and Pesch (2019) stated that 20% of the global container movement and 40–50% of landside movement involves Empty Container Repositioning (ECR). About 56% of the 10–15 years lifespan of the container is also said to usually be spent either being stacked or being repositioned. Shintani et al. (2019) reported that the cost of annual global repositioning of empty containers is estimated at about USD 20 billion. This has prompted research into repositioning, including use of innovations like foldable and combinable containers.

The growth of container shipment has also led to growth in fleet and ship sizes for many carriers in the maritime business. Poo and Yip (2019) noted that the carrying capacity of container ships in the world has increased six-fold from 3.17 million in 1990 to 18.9 million in 2014. In addition, the maximum ship size has progressively increased from 4300 Twenty-Foot Equivalent Units (TEUs) in 1988 to 18,000 TEUs in 2015. Samsung Heavy Industry are currently manufacturing mega ships with capacity to carry 20,000 TEUs. While this development has aided the use of containers for shipment globally, it has further aggravated the imbalance between container supply and demand, thereby necessitating further repositioning.

Lee and Song (2016) classified research areas in container management into six main categories of strategic, tactical, and operational importance with ECR classified mainly as operational but with huge interface with the other areas. They also stated that current research in ECR seeks to answer two inter-related questions of: quantity, which seeks to determine the level of containers to maintain in a location or move between two locations; and cost, which is about how much it costs to reposition containers for subsequent shipments. Within this scope of work, most research seems to have studied capacity deployment, sizes of shipping vessels, design of shipping networks, routing of vessels, and creation of shipping schedules Poo and Yip (2019).

Researchers have reported a general lack of focus on the landside section of container shipping and repositioning with more work done in the long haul or maritime section. Kolar et al. (2018), as part of their findings in a qualitative study, concluded that there has been general negligence of the study of the dynamics of container movement in the hinterland. They surveyed practitioners in the shipping industry in the Central and Eastern Europe (CEE) region and realized that most models proposed hitherto have focused heavily on the maritime side. Sterzik and Kopfer (2013) noted that while the total volume moved inland is much lower than the maritime movement, the unit cost of movement on land is far much higher than the maritime, hence deserves attention in container repositioning. They classified all land movements as either outbound full, outbound empty, inbound full, or inbound empty. They proposed a Mathematical Programming (MP) model for the Inland Container Transportation (ICT) problem considering both resource utilization and container allocation in the hinterland and solved it using Tabu search heuristic. Gusah et al. (2019) also mentioned the sparseness of work in the landside of container repositioning problem and presented an agent-based simulation model of urban-based goods movement in Melbourne, Australia. Furio et al. (2013) is another recent study that has focused mainly on movement of containers considering the implication of street turns as containers are moved inland.

Another area that seems not to have had much work generally in ECR is lot sizing of repositioned containers. Lee and Song (2016) discussed inventory management from container perspective in comparison to that of other products and four main differences were mentioned. Firstly, while for most products the inventory item is consumed, for the container it is more like an equipment used and reused as part of another process. Secondly, while inventory is purchased directly and used for many items, the inventory of containers may be owned, leased, or a combination of both. The third is that container inventory is a 2-way management problem unlike a typical product which is a single-way problem. The fourth difference is consequent to the third, which is that the container system is more like a return logistics system. These characteristics affect how the container inventory system is modeled. The focus is not so much on the price of purchase but the operational cost, and the model is an adaptation of most return logistics inventory models.

The classic model on which most return logistics lot sizing model seems to have been built is the model of Schrady (1967). He presented a model for managing a repairable inventory system in which there are two items: one consisting of items waiting to be repaired, called the Non-Ready for Issue (NRFI) items, and the other containing Ready for Issue (RFI) items. He presented two policies: the continuous supplementing policy and the substitution policy and focussed more on the latter. He considered a case where not all items issued are returned for repair, hence there is a need to procure some new items to supplement the repairable items returned.

He developed a closed form solution that can be used to determine the timing and quantity of lot sizes for each of these two items. Many authors have since modified Schrady's work. A notable one for the context of this paper is Mabini et al. (1992), which presented a multi-item production-return process in which the reparable items share capacity for repair. They also developed models to lot-size the repairable items optimally following the substitution policy of Schrady. Koh et al. (2002) is another relevant work in that they considered that both returned items and new items are mixed and simultaneously consumed because they are similar. This represents how containers behave better than Shrady's in that repaired items are not so distinct from procured items and their consumption cycles are, thus, not necessarily separated during consumption. Another similar work in this wise is that of Cohen et al. (1980) in that returned items can be returned directly to service. In the work presented here, a container return situation is considered where some of the returned containers are put straight back to reuse while some would need to be repaired before reuse and there is also a need to procure some because a proportion is lost. In addition, there are different types of containers in the system, and these different types need to be repaired using a shared facility with finite capacity, thus, planning this repair individually may lead to cycle time overlaps, which is undesirable.

## 11.3 Model Development

This section presents the development of the total cost and optimal lot size functions. The derivation starts by stating the important assumptions of the model, followed by the notations adopted in the development, and then the model derivation.

### *11.3.1 Model Assumptions*

The following assumptions were made in the derivation of the model:

- There are multiple types of containers.
- The monetary value of new stock and repaired items does not affect the operational value and cost of the containers, hence the holding cost for both can be assumed to be the same.
- The demand rate for each type of container is fixed and is known.
- The return rate of each container type is fixed and is known.
- The damage rate for each type of container is fixed and is known.
- The rate of container attrition (loss or irreparably damaged) is much smaller than the rate of return, and consequently, it is reasonable to assume the replenishment cycle is at least as large as the return cycle. This makes it possible to assume that there could be one or more return cycle in a single procurement cycle.

- The demand for RFI containers is fulfilled from all types of sources of RFI (procured, return without repair, or repaired) in a random manner and there is no particular selection preference. This is reasonable because the state of purchase, returned or repair does not confer any advantage on a container. It is selected only for its functionality (serviceability). This makes it possible to assume that they all run out together.

### *11.3.2 Notations*

The following notations were adopted:

$j$ is an index for types of containers managed at the port, $j = 1, 2, \ldots k$
$r$ is a subscript denoting returned containers
$p$ is a subscript denoting purchased containers
$R$ is a subscript denoting repaired containers
$D_j$ is the annual demand rate for container type $j$
$P_j$ is the annual repair rate (repair capacity) for container type $j$
$x_j$ is the proportion of container type $j$ returned to the port from points of use, either in serviceable or repairable form. The proportion $1 - x_j$ either does not return or is returned as being non-serviceable and non-repairable
$y_j$ is the proportion of the returned container type $j$ that needs repair in order to return to serviceable condition after arriving at the port. The proportion $x_j(1 - y_j)$ of the original container issued can be reused on return without any need for any significant repair activity, while the proportion $x_j y_j$ needs to be repaired before it can be reissued for reuse
$K_{rj}$ is the fixed cost of returning a batch of useful container type $j$ from point of use
$K_{pj}$ is the fixed cost of ordering a new batch of container type $j$ to make up for lost or irreparably damaged containers
$K_{Rj}$ is the fixed cost of setting up the repair centre for a batch of container type $j$ to be repaired from the reparable containers contained in the lot of returned containers
$h_j$ is the holding cost per unit per year of keeping a container type $j$. This cost may be dependent only on the type of container but not on the state of repair of the container
$m_j$ is the number of container return cycles per single cycle of procurement for a container type $j$, $m_j \geq 1$ and integer
$n_j$ is the number of container repair cycles per single cycle of return for a container type $j$, $n_j \geq 1$ and integer
$C$ is the aggregate capacity for storage of all the different types of containers in the port terminal
$u_j$ the storage space requirement for a unit of container type $j$
$S_j$ is the time taken to set up the repair centre for a batch of container type $j$ to be repaired from among the batch of returned containers

$t_j$ is the operational time taken to repair a batch of container type $j$, (without considering set up)
$T$ is a common repair cycle time for all containers
$Q^*_{rj}$ is the optimum return quantity for container type $j$
$Q^*_{pj}$ is the optimum replenishment order quantity for container type $j$
$Q^*_{Rj}$ is the optimum repair quantity for container type $j$
$T^*$ is the optimum common repair cycle time for all containers.

### 11.3.3 Model Derivation

Consider a multi-item return container management system discussed (see Introduction). The first focus will be on the dynamics of a single container system out of the $k$ types of containers in circulation, indexed with $j$. The demand for the container per unit time (say per year) is $D_j$. The management arranges for the timely return of the containers to continue to service customers. Only a proportion, $x_j$, of the total number of containers put into circulation is returned to the port where it is needed to be re-used. Of these returned containers, a proportion of the returned container, $y_j$, needs to be repaired before it can be put back into use. After repair, the repaired container is as good as new operationally and can go back to use, such that at the steady-state, the entire proportion $x_j$ of returned containers can be put back to use again. The portion not returned, $(1 - x_j)$, is procured to make up the demand, $D_j$, again. At the repair centre, there is a set up time, $S_j$, necessary to set up the centre for the repair of the container type $j$, and the container can be repaired at a rate $P_j$ per unit time (also say per year). The inventory cost of managing this system comprises of the cost of each of the three subsystems, i.e. purchase, serviceable and reparable units. This is presented next for any container type, $j$.

For the procurement cycle shown in the topmost layer of Fig. 11.2, the total inventory cost rate for management of the container is

$$\frac{(1 - x_j)D_j K_{pj}}{Q_{pj}} + \frac{Q_{pj}h_j}{2} \tag{11.1}$$

For the return container subsystem, the total containers returned is split into two: the portion that can go back into recirculation without repair given by quantity $Q_{rj}(1 - y_j)$, and the portion that needs to be repaired in predetermined batches, $Q_{Rj}$, given by the quantity $Q_{rj}y_j$. The fixed cost for this portion is obvious and is the first term of Eq. 11.2. The holding cost, however, consists of two parts. The first part is the returned container that can go back into circulation without repair, given by the second layer of the diagram from top, and the layer that needs to be repaired and is gradually drawn down in batches of $Q_{Rj}$, the third layer from top. There is a relationship between this third layer and the repair layer (bottom layer). Only the holding cost of the portion not needing repair is represented for now and is given by the second term of Eq. 11.2.

$$\frac{x_j D_j K_{rj}}{Q_{rj}} + \frac{Q_{rj}(1 - y_j)h_j}{2} \tag{11.2}$$

The final (bottom) layer of the Figure is the repair echelon. The fixed repair cost is obvious and would be included later. For the holding cost, the inventory position of the repairable containers in a full return cycle is shown in the two lower levels of Fig. 11.2 (layers 3 and 4 from top). Layer 3 is the position of the reparable containers gradually going into repair drawn in repair batches, and layer 4 is the inventory position of the inducted batches of repair for the container type, $j$. The holding cost for layer 3 is the time weighted average of this layer, i.e. aggregate inventory divided by total time. The proportion of return quantity needing repair, $Q_{rj}y$, is gradually drawn down in repair batches of size $Q_{Rj}$. This forms a pattern in which the first draw is a triangle, followed by a series of trapezia. The height of both the triangle and the trapezia is $Q_{Rj}$. From this, it can be seen that the total inventory position per return cycle for layer 3 (from top) of Fig. 11.2 is

$$\frac{n_j Q_{Rj}}{2P_j} + \frac{Q_{Rj}}{D_j x_j y_j} \sum_{i=1}^{k-1} i \tag{11.3}$$

The cycle time for the return cycle can also be determined by multiplying the number of repair cycles per return cycle, $n_j$, and the cycle time for a single repair, hence,

$$\frac{n_j Q_{Rj}}{D_j x_j y_j} \tag{11.4}$$

Dividing the total inventory per return cycle (Eq. 11.3) by the cycle time (Eq. 11.4), yields the layer's average inventory per unit time, thus

$$\frac{Q_{Rj}h_j}{2}\left[\left(n_j - 1\right) + \frac{D_j x_j y_j}{P_j}\right] \tag{11.5}$$

For the repair layer (layer 4), it can be seen that it is simply an equivalent of a production-inventory system in which there are two phases in a cycle; in the first phase, there is a joint repair and withdrawal of containers and in the second, there is a pure withdrawal period after repair is suspended. The entire cost for the repair subsystem, (layers 3 and 4) can, therefore be written as

$$\frac{x_j y_j D_j K_{Rj}}{Q_{Rj}} + \frac{Q_{Rj}h_j}{2}\left(1 - \frac{D_j x_j y_j}{P_j}\right) + \frac{Q_{Rj}h_j}{2}\left[\left(n_j - 1\right) + \frac{D_j x_j y_j}{P_j}\right] \tag{11.6}$$

Equation 11.6 is the form of the cost function for the repair layers (layers 3 and 4 from the top) for the general repairable container system described in the problem, even when the holding cost is different for RFI and NRFI containers. The first term is the set up cost now included, the second term is the holding cost due to the repaired containers (layer 4) and the third term is the holding cost due to the containers waiting to be repaired (layer 3). If the holding cost of the repaired and repairable terms are considered different, this difference is indicated in that the $h_j$ terms in Eq. 11.6 are differentiated as say $h_{j,1}$ and $h_{j,2}$ where $h_{j,1}$ may be the holding cost rate for the RFI containers and $h_{j,2}$ the holding cost rate for the NRFI containers. In the current derivation, however, it is assumed there is no need for such differentiation for containers and as such, $h_{j,1} = h_{j,2} = h_j$. This makes the cost function simpler by combining the holding cost terms (second and third terms) in Eq. 11.6 which leads to

$$\frac{x_j y_j D_j K_{Rj}}{Q_{Rj}} + \frac{n_j Q_{Rj} h_j}{2} \tag{11.7}$$

One more benefit of the assumption of common holding cost is to be realized later when designing the solution procedure. If the holding cost is the same for both the repairable and repaired containers for a container type and the repair cycle is integrated, the optimum cost is readily found at $n_j = n = 1$.

Adding Eqs. 11.1, 11.2 and 11.7 to derive the total cost rate function for the inventory system yields

$$\frac{(1 - x_j) D_j K_{pj}}{Q_{pj}} + \frac{Q_{pj} h_j}{2} + \frac{x_j D_j K_{rj}}{Q_{rj}} + \frac{Q_{rj}(1 - y_j) h_j}{2} + \frac{x_j y_j D_j K_{Rj}}{Q_{Rj}} + \frac{n_j Q_{Rj} h_j}{2} \tag{11.8}$$

It can still be assumed that there could be one or more return cycles in a single procurement cycle, $m_j \geq 1$, and one or more repair cycles in a single return cycle, $n_j \geq 1$. With this, the relationship between the quantities per batch of procured, returned, and repaired containers relative to the repair cycle time can be written as

$$Q_{pj} = m_j n_j (1 - x_j) D_j T_j \tag{11.9}$$

$$Q_{rj} = x_j n_j D_j T_j \tag{11.10}$$

$$Q_{Rj} = x_j y_j D_j T_j \tag{11.11}$$

Adding Eqs. 11.9–11.13 yields

$$Q_j = D_j T_j n_j \left[ x_j + m_j \left( 1 - x_j \right) \right] \tag{11.12}$$

Substituting Eq. 11.12 into Eq. 11.8 and summing over all container types yields the cost function for the multi-type container system as shown in Eq. 11.13.

$$\sum_{j=1}^{k} \frac{1}{T_j}\left[\frac{K_{pj}}{m_j n_j} + \frac{K_{rj}}{n_j} + K_{Rj}\right] + \sum_{j=1}^{k}\left[\frac{D_j T_j h_j n_j}{2}\left[m_j(1 - x_j) + x_j\right]\right] \quad (11.13)$$

Since all containers share a common repair facility that needs to be set up for the repair of each type of container, it is important that the repair of each of these types of containers be completed within a repair cycle. All types of container varieties are catered for in this cycle, and this is a condition for the feasibility of the solution as it prevents cycle overlaps. It then becomes necessary to replace the individual cycle times for each container, $T_j$, by a common cycle time, $T$. If the cycle times are integrated this way, it also becomes pertinent that the individual number of repair cycles per return cycle,$n_j$, be replaced by a common number of repairs, $n$. Equation 11.13, therefore, becomes

$$\frac{1}{T}\sum_{j=1}^{k}\left[\frac{K_{pj}}{m_j n} + \frac{K_{rj}}{n} + K_{Rj}\right] + \frac{Tn}{2}\sum_{j=1}^{k}\left[D_j h_j\left[m_j(1 - x_j) + x_j\right]\right] \quad (11.14)$$

The diagram for the common repair cycle for all container types is shown in Fig. 11.3. For this cycle to be feasible, there is the constraint that all setups and repairs for all containers must be completed within each repair cycle time, $T$. This is represented as

$$\sum_j S_j + \sum_j t_j \le T \quad (11.15)$$

The operational repair time (i.e. excluding set up time) for a repair batch of container $j$, $t_j$, can be expressed as

$$t_j = \frac{Q_{Rj}}{P_j} \quad (11.16)$$

Also define, $\rho_j$, the utilization level for the repair resource based on container $j$ to be

$$\rho_j = \frac{D_j}{P_j} \quad (11.17)$$

Using Eq. 11.16 with Eqs. 11.11 and 11.17, one can rewrite Eq. 11.15 as

$$T \ge \frac{\sum_j S_j}{1 - \sum_j x_j y_j \rho_j} \quad (11.18)$$

There is also the possibility of having insufficient space to store the containers at the port. The quantity of return containers that is repositioned in addition to the quantity purchased and the current containers under repair could be limited by the storage space and not the capacity to repair the damaged returned containers. This constraint can be expressed as

$$\sum_{j=1}^{k} Q_j u_j \leq C \tag{11.19}$$

Substituting Eq. 11.12 for $Q_j$ in Eq. 11.19 with $T_j = T \, \forall \, j$ yields

$$T \leq \frac{C}{n \sum_{j=1}^{n} D_j \big[ (x_j + m_j (1 - x_j) \big] u_j} \tag{11.20}$$

The general problem becomes that of minimizing the cost function, Eq. 11.14, subject to the two constraints, Eqs. 11.18 and 11.20. In solving this problem, it can be seen that Eqs. 11.18 and 11.20 represent the upper and lower bounds for any feasible cycle time. This becomes the approach to exploit in solving the problem and would be discussed later.

Optimizing Eq. (11.14) with respect to $T$ while ignoring the constraint yields

$$T^* = \sqrt{\frac{2 \sum_{j=1}^{k} \left[ \frac{K_{pj}}{m_j n} + \frac{K_{rj}}{n} + K_{Rj} \right]}{n \sum_{j=1}^{k} \big[ D_j h_j \big[ m_j (1 - x_j) + x_j \big] \big]}} \tag{11.21}$$

To solve Eq. 11.21, it should be observed that $m_j$ and $n$ values are also unknown, and the problem would have to be solved iteratively. This can be solved through some numerical approaches or some other search techniques, including random search solutions. Once the values of $m_j$ s and $n$ that optimize the cycle time have been determined for each of the container types, one needs to check the solution for feasibility using Eqs. 11.18 and 11.20. If the constraints are violated, one needs to determine the new cycle time that would be feasible based on either Eq. 11.18 or Eq. 11.20, depending on which one is violated. Once the cycle time is deemed acceptable, the optimal lot size for each of the containers for the return, purchase, and repair lots can be calculated from Eqs. 11.9 to 11.11.

Alternatively, to include the impact of the two constraints on the objective function, one can create a Lagrangian function including the two constraints and solve. This leads to a series of simultaneous equations involving the differentiation of the Lagrangian with respect to the cycle time for each $j$, with respect to the first and second Lagrange variables, and with respect to the common number of repairs per return cycle. This may be computationally tedious, and even so, there would still be the need to iterate over $m_j$.

### 11.3.4 Proof of Optimality

To check if the cycle time determined in Eq. 11.21 minimizes the cost functions, it suffices to check the hessian function of Eq. 11.15, which is given by Eq. 11.22. It can be seen that this function is positive definite since all terms are positive non-zero values. It can, therefore, be concluded that Eq. 11.21 provides a minimum for Eq. 11.14.

$$\frac{1}{T^3}\sum_{j=1}^{k}\left[\frac{K_{pj}}{m_j n}+\frac{K_{rj}}{m_j}+K_{Rj}\right] \tag{11.22}$$

## 11.4 Solution Algorithm

The solution procedure iteratively finds the best feasible solution within the search space until a better one could not be found, or it finds a reasonable infeasible solution as a bound on the cost value and from there seeks out a close feasible solution on the constraint boundary (Fig. 11.4).

The solution exploits the fact that all repaired items must have a common cycle time so that every container type is repaired during each repair cycle, hence, only needs to iterate over a single $n$ for all containers and not $n_j$ for each container $j$ as discussed earlier. It starts by fixing the value of $n$ at 1 and uses the solver to search for the best combinations of $m_j$ at $n = 1$. It proceeds to $n = 2$ and iterates. If a better result is obtained, the solution at $n = 2$ is kept as the best. The iteration continues until the solution starts to deteriorate or becomes infeasible. If the best was found before it becomes infeasible, then it keeps the best feasible solution it has found. If it moves to an infeasible region before it starts to deteriorates, there is the need to check for the better between the last feasible solution found and the boundary solution close to the infeasible region. To find the boundary solution close to the infeasible region, the $n$ and $m_j$ combinations obtained as the last feasible solution is used with the repair cycle time set to the boundary cycle time $T$ determined from Eqs. 11.18 or 11.20, depending on which one led to the infeasibility of the solution. The choice of the best cycle time and number of repair cycle per return cycle is made based on the better of these last two candidate solutions.

## 11.5 Numerical Examples

Two numerical examples were proposed to demonstrate the use of the algorithm to find solution for the problem. The first problem (numerical example) is a 3-container problem and the second is a 6-container problem. The storage capacity

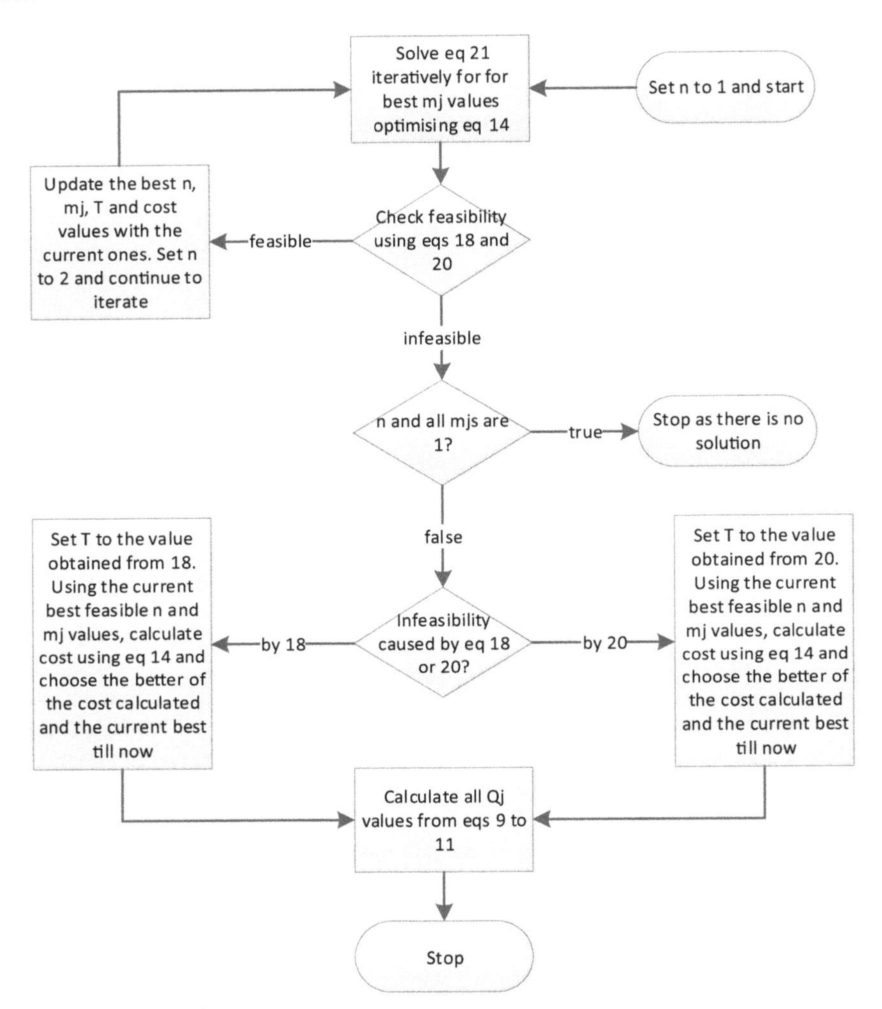


**Fig. 11.4** Solution algorithm flow chart

are $C = 350000\,ft^3$ and $C = 2000000\,ft^3$ respectively for the 3- and 6-container problems while the other problem parameters are also presented in Tables 11.1 and 11.2 respectively. The optimal solutions for these two problems are presented in Table 11.3 with $n = 1$ as part of the solution. This $n$ value is understandable since the holding cost, repair cycle time and the number of repair cycle per return cycles have been standardized. Sensitivity analysis was done for the result obtained for the 6-container problem only as shown in Table 11.4. This is without any loss of generality, but only to avoid unnecessary repetition.

The following observations from the sensitivity analysis are noteworthy:

**Table 11.1** Input parameters for the 3-container problem

| $j$ | 1 | 2 | 3 |
|---|---|---|---|
| $u_j(ft^3/containers)$ | 20 | 15 | 10 |
| $D_j(containers/year)$ | 150 000 | 20000 | 25000 |
| $P_j(containers/year)$ | 18 000 | 24 000 | 30 000 |
| $x_j$ | 0.9 | 0.8 | 0.8 |
| $y_j$ | 0.2 | 0.3 | 0.25 |
| $K_{rj}$ ($) | 10 000 | 8000 | 7000 |
| $K_{pj}$ ($) | 15 000 | 12 000 | 10 500 |
| $K_{Rj}$ ($) | 20 000 | 16 000 | 14 000 |
| $h_j(\$/year)$ | 50 | 40 | 30 |
| $S_j(years)$ | 0.02 | 0.018 | 0.016 |

**Table 11.2** Input parameters for the 6-container problem

| $j$ | 1 | 2 | 3 | 4 | 5 | 6 |
|---|---|---|---|---|---|---|
| $u_j(ft^3/containers)$ | 50 | 35 | 25 | 20 | 15 | 10 |
| $D_j(containers/year)$ | 1000 | 5000 | 10 000 | 100 000 | 25 000 | 50 000 |
| $P_j(containers/year)$ | 1250 | 6250 | 12 500 | 120 000 | 30 000 | 60 000 |
| $x_j$ | 0.9 | 0.8 | 0.75 | 0.75 | 0.7 | 0.85 |
| $y_j$ | 0.15 | 0.16 | 0.18 | 0.15 | 0.17 | 0.2 |
| $K_{rj}$ ($) | 10 000 | 7000 | 5000 | 4000 | 3000 | 2000 |
| $K_{pj}$ ($) | 15 000 | 12 000 | 11 250 | 10 500 | 9750 | 9000 |
| $K_{Rj}$ ($) | 20 000 | 15 000 | 13 000 | 8000 | 7000 | 5000 |
| $h_j(\$/year)$ | 50 | 40 | 30 | 20 | 10 | 5 |
| $S_j(years)$ | 0.015 | 0.0135 | 0.012 | 0.0105 | 0.009 | 0.0075 |

**Table 11.3** Solution to the 3- and 6-container problems

| | Total cost ($TC$) | Common cycle time ($T$) | Number of return cycles per procurement cycle ($m_j$) |
|---|---|---|---|
| Example 1 (3 containers) | 706 222.34 | 0.2973 | 2; 1;1 |
| Example 2 (6 containers) | 951 619.67 | 0.2937 | 8; 3; 2; 1; 1;1 |

- For all the different parameter settings tested the number of repair cycles per return cycle ($n$) remained constant at 1 as expected.
- None of the changes tested with respect to the storage capacity for all container types ($C$) had any effects on the total inventory management cost ($TC$), the number of return cycles per procurement cycle ($m_j$) and the common cycle time ($T$). This is because in the example tested, the storage capacity constraint, given in Eq.

**Table 11.4** Sensitivity analysis for the 6-container problem

| Parameters | % change | Total cost ($TC$) | % change | Common cycle time ($T$) | % change | Number of return cycles per procurement cycle ($m_j$) |
|---|---|---|---|---|---|---|
| $K_{rj}$ | −50 | 897 234.87 | −5.71 | 0.2765 | −5.86 | 9; 3; 2; 1; 1;1 |
| | −25 | 924 839.40 | −2.81 | 0.2850 | −2.96 | 9; 3; 2; 1; 1;1 |
| | +25 | 977 650.24 | +2.74 | 0.3017 | +2.74 | 8; 3; 2; 1; 1;1 |
| | +50 | 1 003 005.48 | +5.40 | 0.3096 | +5.40 | 8; 3; 2; 1; 1;1 |
| $K_{pj}$ | −50 | 877 157.34 | −7.82 | 0.2816 | −4.13 | 6; 3; 2; 1; 1;1 |
| | −25 | 915 845.55 | −3.76 | 0.2866 | −2.40 | 7; 3; 2; 1; 1;1 |
| | +25 | 985 598.43 | +3.57 | 0.3037 | +3.41 | 9; 3; 2; 1; 1;1 |
| | +50 | 1 018 408.69 | +7.02 | 0.3134 | +6.69 | 10; 3; 2; 1; 1;1 |
| $K_{Rj}$ | −50 | 827 609.51 | −13.03 | 0.2547 | −13.30 | 10; 3; 2; 1; 1;1 |
| | −25 | 891 793.37 | −6.29 | 0.2748 | −6.43 | 9; 3; 2; 1; 1;1 |
| | +25 | 1 007 839.27 | +5.29 | 0.3111 | +5.91 | 8; 3; 2; 1; 1;1 |
| | +50 | 1 060 544.25 | +11.47 | 0.3319 | +13.02 | 7; 3; 2; 1; 1;1 |
| $h_j$ | −50 | 672 896.72 | −29.29 | 0.4153 | +41.42 | 8; 3; 2; 1; 1;1 |
| | −25 | 824 126.81 | −13.40 | 0.3391 | +15.47 | 8; 3; 2; 1; 1;1 |
| | +25 | 1 063 943.14 | +11.80 | 0.2627 | −10.56 | 8; 3; 2; 1; 1;1 |
| | +50 | 1 165 491.31 | +22.47 | 0.2398 | −18.57 | 8; 3; 2; 1; 1;1 |
| $S_j$ | −50 | 951 619.67 | 0 | 0.2937 | 0 | 8; 3; 2; 1; 1;1 |
| | −25 | 951 619.67 | 0 | 0.2937 | 0 | 8; 3; 2; 1; 1;1 |
| | +25 | 951 619.67 | 0 | 0.2937 | 0 | 8; 3; 2; 1; 1;1 |
| | +50 | 951 619.67 | 0 | 0.2937 | 0 | 8; 3; 2; 1; 1;1 |
| $u_j$ | −50 | 951 619.67 | 0 | 0.2937 | 0 | 8; 3; 2; 1; 1;1 |
| | −25 | 951 619.67 | 0 | 0.2937 | 0 | 8; 3; 2; 1; 1;1 |
| | +25 | 951 619.67 | 0 | 0.2937 | 0 | 8; 3; 2; 1; 1;1 |
| | +50 | 951 619.67 | 0 | 0.2937 | 0 | 8; 3; 2; 1; 1;1 |
| $C$ | −50 | 951 619.67 | 0 | 0.2937 | 0 | 8; 3; 2; 1; 1;1 |
| | −25 | 951 619.67 | 0 | 0.2937 | 0 | 8; 3; 2; 1; 1;1 |
| | +25 | 951 619.67 | 0 | 0.2937 | 0 | 8; 3; 2; 1; 1;1 |
| | +50 | 951 619.67 | 0 | 0.2937 | 0 | 8; 3; 2; 1; 1;1 |

(11.20), was not violated by any of the percentage changes yet. Consequently, the solution remained intact because $C$ has no direct effect on $TC$ and $T$ when the storage capacity constraint is not violated since the optimal values are determined using Eqs. (11.15) and (11.21) respectively.

- Changing the holding costs ($h_j$) had significant effects on both $TC$ and $T$ but not on $m_j$. As expected, $TC$ increased with increasing holding costs and $T$ decreased with increasing holding costs. That the $m_j$ values are not significantly affected by changes to the holding cost is likely because $(1 - x_j)$ is small and nuances the effects with the changes on $m_j$ except when such changes are very significant.
- Changes to both storage space requirement for a single container type ($u_j$) and the duration of time required to set up a repair centre ($S_j$) did not have any effects on $m_j$ and $T$. Similar to $C$, both parameters (i.e. $u_j$ and $S_j$) have no direct effect on the expressions for $TC$ and $T$, as given in Eqs. (11.15) and (11.21), when the storage capacity constraint (in the case of $u_j$) and the repair time constraint (in the case of $S_j$) are not violated. They only start affecting the optimal solution when those constraints are violated and for this particular example, none of the percentage changes tested resulted in a violation of the two constraints.
- Changes to all three fixed costs (i.e.$K_{rj}$, $K_{pj}$ and$K_{Rj}$) resulted in significant effects on $TC$, $m_j$ and $T$. The resulting effects followed the same general pattern, the total cost and the cycle time increase with increasing fixed costs. This result is not surprising because the objective of the solution is to minimize costs and if the fixed costs are increased, the solution responds by reducing the number of setups (or orders placed) and this is achieved by increasing the order quantity (which means fewer setups) and thus the cycle time is increased as well. While all three fixed costs showed the same general response pattern, the degree of sensitivity differed among them, with $K_{Rj}$ being the most sensitive and $K_{rj}$ being the least sensitive parameter among the three. $K_{rj}$ is expected to be the most sensitive because it is neither divided by $n$ nor$m_j$, while $K_{pj}$ should have been the least sensitive if $n$ is greater than 1. The effect of $n$ dividing $K_{pj}$ is however not really seen on $K_{pj}$ because $n = 1$.

## 11.6 Conclusion

A model of multi-type container return management in which some of the containers are repaired in a facility with shared repair capacity and limited storage capacity was presented. The cost function for the economic quantities to purchase, collect and repair were derived for the joint return system and the constraining equations for the cycle times (upper and lower bounds) based on repair and storage capacities were derived. The lot sizing functions cannot be solved in closed form, hence, an algorithmic solution was proposed. The algorithm iteratively seeks the optimal combination of return numbers for a given number of repair until a turning point or the first point of infeasibility is obtained. From this, either the latest feasible solution observed or the boundary region solution near the infeasible solution is selected as the optimal. The model and solution approach would be useful for a port management authority seeking to optimize the cost of managing empty containers in a complex environment.

## References


Cohen MA, Pierskalla WP, Nahmias S (1980) A dynamic inventory system with recycling. Naval Res Logist Quart 27(1):289–296

Furió S, Andrés C, Adenso-Díaz B, Lozano S (2013) Optimization of empty container movements using street-turn: application to Valencia hinterland. Comput Ind Eng 66:909–917

Gencer H (2019) An overview of empty container repositioning studies and research opportunities, business and management horizons, vol 7, No 1

Gusah L, Cameron-Rogers R, Thompson RG (2019) A systems analysis of empty container logistics—a case study of Melbourne Australia. Transp Res Proc 39:92–103

Kolar P, Schramm H-J, Prockl G (2018) Intermodal transport and repositioning of empty containers in Central and Eastern Europe hinterland. J Transp Geogr 69:73–82

Kuzmicz KA, Pesch E (2019) Approaches to empty container repositioning problems in the context of Eurasian intermodal transportation. Omega 85:194–213

Koh S-G, Hwang H, Sohn K-I, Ko C-S (2002) An optimal ordering and recovery policy for reusable items. Comput Ind Eng 43:59–73

Lee C-Y, Song D-P (2016) Ocean container transport in global supply chains: overview and research opportunities. Transp Res Part B 95:442–474

Levinson M (2016) The box: how the shipping container made the world smaller and the world economy bigger, 2 ed. Princeton University Press

Mabini M, Pintelon L, Gelders L (1992) EOQ type formulations for controlling repairable inventories. Int J Prod Econ 28(2):1–33

Poo MC-P, Yip TL (2019) An optimization model for container inventory management. Ann Oper Res 273:433–453

Rodrigue JP, Notteboom T (2009) The geography of containerization: half a century of revolution, adaptation and diffusion. GeoJournal 74:1

Schrady DA (1967) A deterministic inventory model for reparable items. Naval Res Logist Quart 14(3):391–398

Shintani K, Konings R, Imai A (2019) Combinable containers: a container innovation to save container fleet and empty container repositioning costs. Transp Res Part E 130:248–272

Sterzik S, Kopfer H (2013) A Tabu search heuristic for the inland container transportation problem. Comput Oper Res 40(2013):953–962

Vojdani N, Lootz F, Rösner R (2013) Optimizing empty container logistics based on a collaborative network approach. Maritime Econ Logist 15(4):467–493

Yu M, Fransoo JC, Lee C-Y (2018) Detention decisions for empty containers in the hinterland transportation system. Transp Res Part B 110:188–208

# Chapter 12
# Inventory Management Under Carbon Emission Policies: A Systematic Literature Review

**Arash Sepehri**

**Abstract** As the emission of carbon dioxide has resulted in many issues in the global environment, controlling carbon emission has become a high priority for governments. One of the sectors engaged with carbon emission is inventory management. A lot of activities in inventory systems such as purchasing, warehousing, and transporting the items lead to emitting carbon. Therefore, governments have ruled policies to mitigate the emissions in inventory systems and develop sustainable supply chains. Despite the importance of this issue, no attempts have been made to study and address the vital role of different policies in controlling carbon emissions in review progress. This paper provides a systematic literature review to analyze the impact of carbon emission policies on inventory systems. 75 papers have been extracted from the most relevant academic and research databases and the results have been analyzed and synthesized. By classifying and introducing different carbon policies applicable in inventory systems, this paper introduces the policies that make effort to restrict the emissions. Finally, theoretical and managerial insights and extensive opportunities for future research are outlined.

**Keywords** Inventory management · Carbon emission policy · Systematic literature review

## 12.1 Introduction

The terms greening or environmental consideration in the business refer to regulating the environmental considerations for the stakeholders to satisfy the expectations and decrease relevant costs (Gupta 1995). In addition, industrialization has led to increases in demand for resources and leads to an increase in human-caused carbon emissions (Panayotou 1993; Jia et al. 2020). Therefore, a conflict between sustainability and economic growth is raised (Gupta 1995). In December 1997, the adoption of the Kyoto protocol became an important achievement to adjust the carbon emissions to avoid their impact on climate changes (Oberthür and Ott

A. Sepehri (✉)
School of Industrial Engineering, Iran University of Science and Technology, Tehran, Iran



N. H. Shah et al. (eds.), *Decision Making in Inventory Management*,
Inventory Optimization, https://doi.org/10.1007/978-981-16-1729-4_12

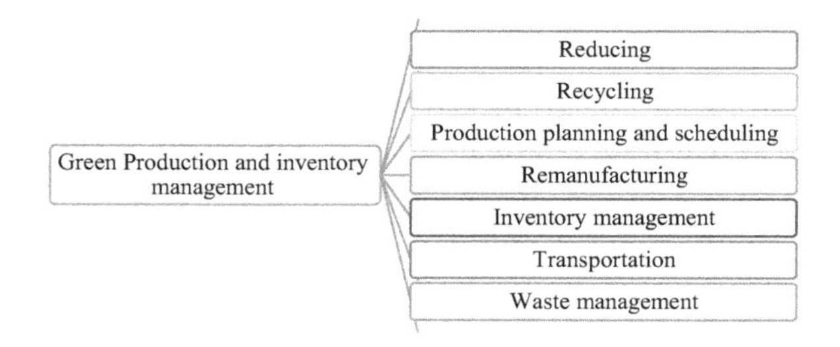


**Fig. 12.1** Green operations in inventory and production systems. *Source* own

1999). The term "sustainable development" indicates a contribution of social and economic considerations along with environmental ones (Pope et al. 2004). Operations management is mainly concentrated on achieving a trade-off between sustainability considerations and the profit obtained (Kleindorfer et al. 2005). Supply chains are emphasizing sustainability considerations from the initial steps in the supply chain such as purchasing raw materials to the final processes such as customer delivery (Linton et al. 2007; Seuring and Müller 2008; Bansal et al. 2018; Iqbal and Sarkar 2020,2019). In addition, green operations are engaged with inventory and production systems and they relate to all aspects of an inventory or a production system such as manufacturing, remanufacturing, transporting, usage, handling, and waste management (Srivastava 2007). Green operations in inventory and production systems are classified in Fig. 12.1.

One of the most challenging environmental concerns in inventory management is greenhouse gases (GHG) emission and carbon dioxide is one of the major gases emit due to inventory and production operations which lead to an increase in global warming (Bonney and Jaber 2011). As different collaborators are engaged in an inventory system, it won't be plausible to mitigate the emissions from the entire inventory system without coordination (Benjaafar et al. 2012). Moreover, concerns from the consumers about carbon dioxide emissions have been increased in recent years and their tendency to purchase the items from manufacturers that produce sustainable products has been intensified (Walker et al. 2014). Therefore, the firms concentrate on identifying a trade-off between environmental considerations and the costs related to inventory systems (Dekker et al. 2012). In this regard, various policies are regulated to control the emissions of greenhouse gases by restricting or encouraging manufacturers (Wee and Daryanto 2020).

Despite the importance of carbon emission regulations in inventory systems, no research has been made to address the importance of carbon emissions in inventory systems directly in the context of a literature review. To fill this gap, this paper proposes a systematic literature review on the impact of carbon emission considerations on inventory management. The relevant papers are exposed using the research methodology presented in Sect. 12.2. Section 12.3 provides statistical analysis for the papers identified based on classifying them into different categories. Section 12.4 discusses carbon emission policies and the employment of each policy in inventory models. Then, the findings are discussed and summarized in a table to categorize the papers is provided in Sect. 12.5. Finally, theoretical and managerial contributions and research gaps recognized in our study are provided in Sect. 12.6.

## 12.2 Research Methodology

In this paper, a systematic literature review (SLR) is applied to collect, synthesize, and analyze the relevant papers. In order to carry out the SLR, (1) suitable search keywords are defined, (2) relevant literature on the topic is identified, and (3) analysis on the literature is performed (Kamble et al. 2018). Numerous relevant papers have been worked on the impacts of carbon emissions on inventory systems. To have access to a wide variety of papers and present an appropriate analysis, papers from the Google Scholar, Scopus, and Web of Science databases were selected. Papers mainly from renowned publications were synthesized and reviewed (Nguyen et al. 2018). The structure below is based on the three-step structure presented by Tranfield et al. (Tranfield et al. 2003) and developed for our elaboration (see Fig. 12.2).

In the first step, we have chosen Scopus, Google Scholar, and Web of Science as our main databases to find the papers. In the second step, we mainly defined the most relevant keywords applicable to our topic to provide an unbiased paper search process. The relevant keywords selected are divided into two distinct categories:

**Category 1**: *Inventory management, Inventory control, Lot sizing, Economic order quantity, Economic production quantity, Production management, Production control.*

**Category 2**: *Carbon emission policy, Carbon emission regulation, Carbon trade, Carbon cap, Carbon tax, Carbon offset, Carbon investment, Sustainability.*

The paper selection in the third step is performed using one keyword from each category and then the contribution of different words from both categories at a time.

In the fourth step, duplicate papers that were found in more than one contribution of two keywords from both categories were eliminated. Moreover, the relevancy of the

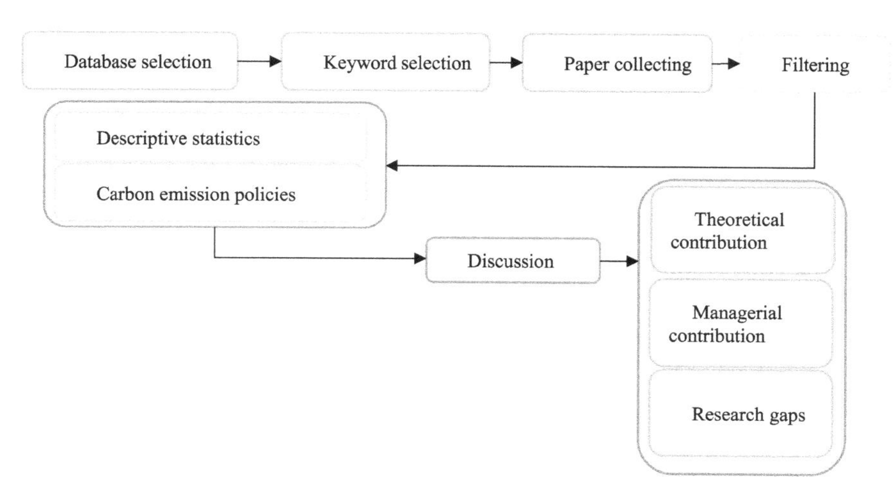


**Fig. 12.2** Systematic literature review. *Source* own

papers to the topic is analyzed and the unrelated models to inventory management are removed from the papers. In the fifth step, we provided a discussion on the statistics and we defined different carbon emission policies and their applications. We proposed a summary of the classification of the papers and we concluded the discussion in the seventh step. Finally, 75 papers were selected from the searching progress.

## 12.3 Descriptive Statistics

### *12.3.1 Year Wise Publications*

In order to illustrate the trend of the number of papers published in recent years, we studied papers from (2011–2020) and we divided this period into two-year intervals (see Fig. 12.3). The number of papers published is dramatically increased in the last ten years. A considerable upward trend is shown in all intervals. In conclusion, the increasing trend in the number of papers published shows the importance of this subject.

### *12.3.2 Contributions by Journals*

The contribution made by different journals is analyzed. Journal of Cleaner Production has the highest number of publications with 28 papers and followed by the International Journal of Production Economics with 10 papers. A summary of this contribution is exposed in Fig. 12.4.

The authors contributed to the papers reviewed in this study are extracted from the reviewing process. The most important researchers with the highest numbers of papers they have contributed to them in this study are shown in Fig. 12.5.

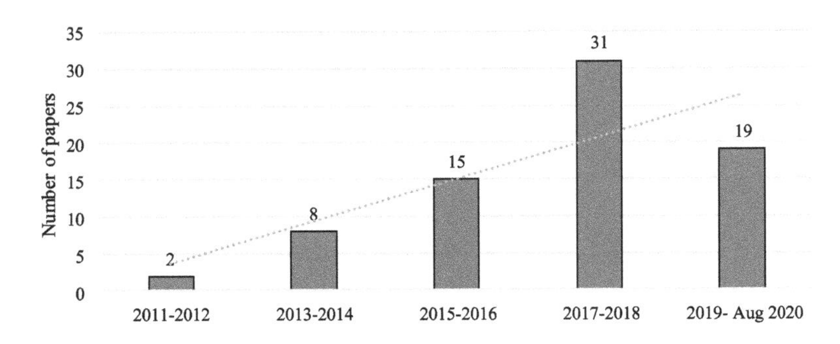


**Fig. 12.3** Year wise publications. *Source* own

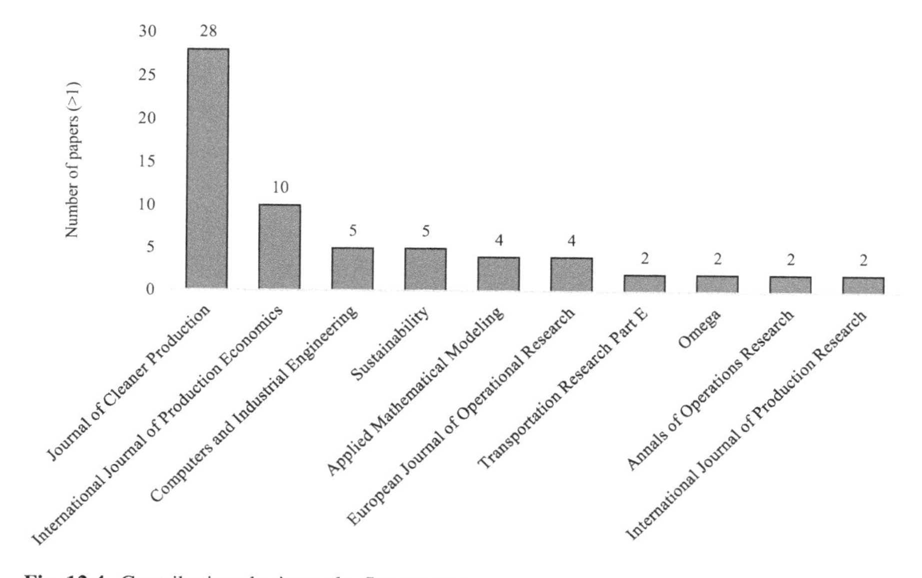


**Fig. 12.4** Contributions by journals. *Source* own

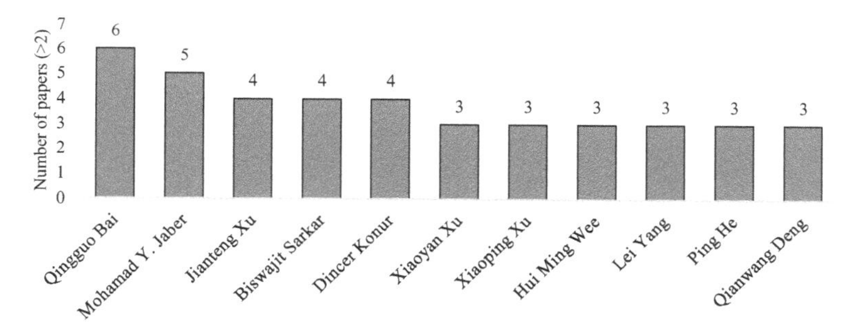


**Fig. 12.5** Contributions by authors. *Source* own

## 12.3.3 Contributions by Countries

The authors' affiliations with different countries are extracted and it is illustrated that China and Taiwan are the countries with the highest paper contribution. Canada and the United States are next on this list. An increase in the variety of countries considers carbon emission reduction in inventory systems is exposed (see Fig. 12.6).

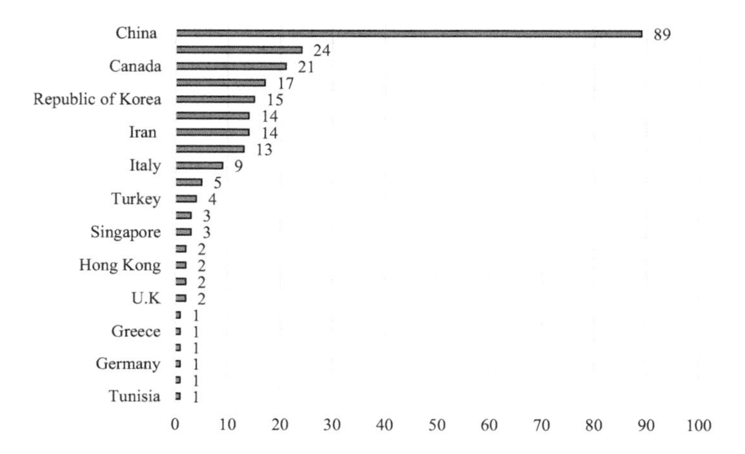


**Fig. 12.6** Contributions by countries. *Source* own

### *12.3.4 Distribution of Carbon Emission Causes*

Carbon emissions take place because of ordering, warehousing, purchasing, setup, transportation, and other operations and events related to inventory management. Warehousing is the most important cause of emitting carbon in inventory systems. The summary of the distribution of the most important causes of carbon emission is exposed in Fig. 12.7.

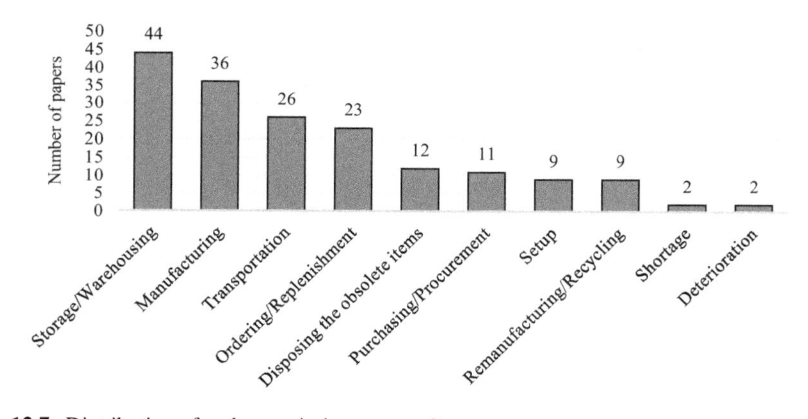


**Fig. 12.7** Distribution of carbon emission causes. *Source* own

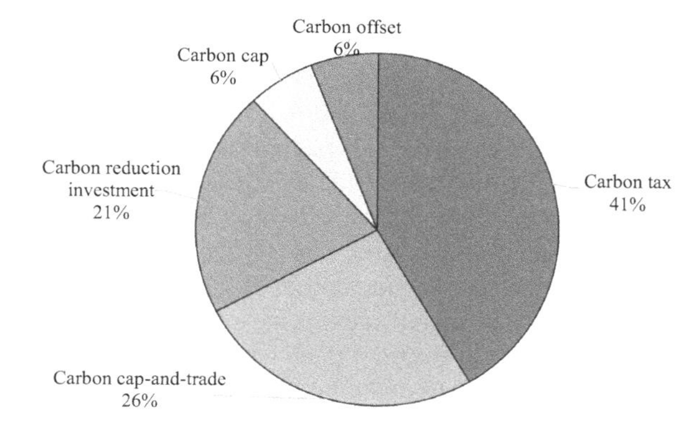


**Fig. 12.8** Distribution of carbon emission policies. *Source* own

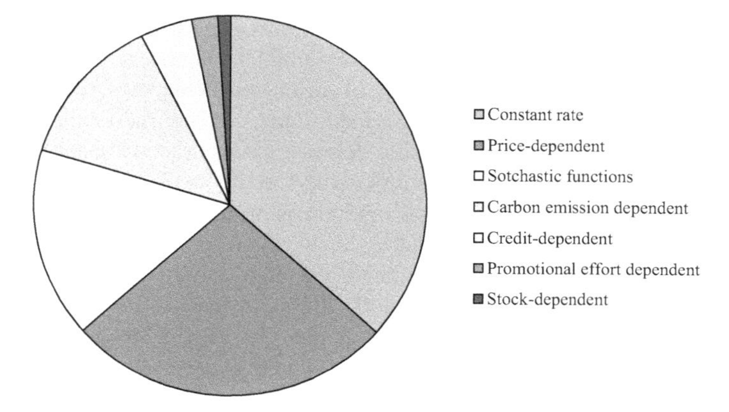


**Fig. 12.9** Distribution of demand functions. *Source* own

### 12.3.5 Distribution of Carbon Emission Policies

Carbon tax and carbon cap-and-trade are two important policies that can have a significant influence on mitigating carbon emissions. The most important policies utilized to mitigate the carbon emissions in inventory systems are shown in Fig. 12.8.

### 12.3.6 Distribution of Demand Functions

The most important demand functions utilized in inventory models reviewed in this research are shown in Fig. 12.9. The most frequent demand function is the constant rate demand and the second frequent demand function is price-dependent demand.

## 12.4 Literature Review

Hua et al. (2011) first took carbon emission into consideration in inventory models under carbon cap-and-trade regulation. This paper was elaborated by Bouchery et al. (2012) and Hovelaque and Bironneau (2015) by considering carbon tax policy and Arslan and Turkay (Hua et al. 2011) by considering carbon offset policy. The main difference between (Bouchery et al. 2012) and (Hovelaque and Bironneau 2015) is the carbon emission-dependent demand considered by Hovelaque and Bironneau (2015). Chen et al. (2013) considered purchasing operation as another source of emissions. Jaber et al. (2013) developed a model and considered manufacturing as the main source of carbon emissions and emissions can be reduced using reduction technologies. Their work was extended by Krass et al. (2013) and Lou et al. (2015) by considering a price-dependent demand. Transportation and shortage are other sources of emissions (Bazan et al. 2015b; Konur 2014; Battini et al. 2014; Bozorgi et al. 2014; Gurtu et al. 2015). Toptal et al. (2014) extended (Jaber et al. 2013) considering storage, holding, and purchasing inventory as main sources of emissions simultaneously. Alongside manufacturing, remanufacturing (Bazan et al. 2015a) and setup (He et al. 2015) can be another cause of emissions. Chang et al. (2015) proposed a manufacturing and remanufacturing model under carbon reduction investment. Dye and Yang (2015) added permissible delay in payment to sustainable inventory models considering a credit-dependent demand and this work was followed by Qin et al. (2015) which developed a credit and emission-dependent demand. Schaefer and Konur (2015) developed a model under uncertain demand which is followed by Chen et al. (2016). Some papers considered other factors as the main causes of emission such as deterioration (Rout et al. 2020; Huang et al. 2018) and shortage (Konur et al. 2017; Schaefer and Konur 2015). Developing models for defective items is another novelty considered in many papers (Rout et al. 2020; Sarkar et al. 2016, 2018; Datta 2017; Shu et al. 2017; Kazemi et al. 2018; Tiwari et al. 2018). Afterward, many papers developed inventory models under carbon cap-and-trade (Lu et al. 2020; Liao and Deng 2018a; Battini et al. 2018; Mishra et al. 2020b; Halat and Hafezalkotob 2019; Rout et al. 2020; Cao et al. 2017; Dong et al. 2016; Xu et al. 2016a, b,2017, 2018; Marklund and Berling 2017; Sabzevar et al. 2017; Shu et al. 2017; Tsao et al. 2017; Wang et al. 2017; Yang et al. 2017; Aljazzar et al. 2018; Bai et al. 2018, 2019; Turki et al. 2018; Heydari and Mirzajani 2020), carbon tax (Konur et al. 2017; Ji et al. 2017; Mishra et al. 2020a; Tang et al. 2018; Rout et al. 2020; Sarkar et al. 2016, 2018a, b; Dong et al. 2016; Xu et al. 2016b; Bouchery et al. 2017; Datta 2017; Tsao et al. 2017; Aljazzar et al. 2018; Darom et al. 2018; Huang et al. 2018; Kazemi et al. 2018; Liao and Deng 2018b; Lin 2018; Taleizadeh et al. 2018, 2020; Tiwari et al. 2018; Zhou et al. 2018; Aliabadi et al. 2019; Daryanto et al. 2019; Jabbarzadeh et al. 2019; Li and Hai 2019; Shi et al. 2019, 2020; Bai et al. 2020; Datta et al. 2020; Wangsa et al. 2020), carbon offset (Lu et al. 2020; Tang et al. 2018; Halat and Hafezalkotob 2019; Rout et al. 2020), and carbon reduction investment (Ji et al. 2017; Lu et al. 2020; Mishra et al. 2020a,b; Battini et al. 2018; Tang et al. 2018; Sarkar et al. 2016; Cao et al. 2017; Dong et al. 2016; Xu et al. 2016a; Datta

2017; Yang et al. 2017; Bai et al. 2018, 2019,2020; Shi et al. 2019; Datta et al. 2020; Kang et al. 2020; Wang and Hui 2020).

## 12.5 Carbon Emission Policies

As we discussed in previous sections, this paper aims to study different carbon emission regulations in inventory management. Different policies are discussed as follows.

### *12.5.1 Carbon Cap*

For firms that have high carbon emissions in their operations, a carbon cap is allocated by the government to reduce emissions as much as possible (Chen et al. 2013; Bazan et al. 2015b). Carbon cap policy usually exposes as a constraint in an inventory model that restricts the total carbon emissions to a specific cap (Konur et al. 2017). The objective is to find the minimum total inventory costs (Konur et al. 2017; Konur 2014; Schaefer and Konur 2015; Ghosh et al. 2017) or to find the maximum total profit (Qi et al. 2017). An illustration of the carbon cap policy is provided as follows.

$$\begin{aligned} \text{Objective: } & \max \text{Total profit} \\ \text{or: } & \min \text{Total cost} \\ \text{Subject to: } & CE \le Z \end{aligned} \tag{12.1}$$

where $CE$ is the total carbon emissions and $Z$ is the carbon cap.

### *12.5.2 Carbon Tax*

The carbon tax is one of the most frequent policies regulated by the government to mitigate carbon emissions (Ghosh et al. 2017; Bouchery et al. 2012). A tax is considered for the total carbon emissions that genera9te a carbon emission cost (Jaber et al. 2013; Hovelaque and Bironneau 2015; Ji et al. 2017). An indication of the carbon tax policy is proposed as follows

$$\delta(CE) \tag{12.2}$$

where $\delta$ is the carbon tax and $CE$ is total carbon emissions.

### 12.5.3 Carbon Cap-And-Trade

Carbon cap-and-trade is an encouraging policy regulated by the governments for the firms (Hua et al. 2011; Arslan and Turkay 2013). This policy considers a carbon cap for the firms and if the firms' emissions stand less than the carbon cap, they can buy the surplus of the cap and accumulate revenue (Lu et al. 2020; Mishra et al. 2020a). This policy allows the firms to earn more profit due to restricting their carbon emissions (Liao and Deng 2018a; Toptal et al. 2014). The income from the selling of surplus carbon allowance will be calculated with the carbon tax rate (Battini et al. 2018; Mishra et al. 2020b). Then, the revenue obtained will be added to the objective function if the optimal profit is the objective or it will be subtracted from the objective function if the optimal cost is the objective. An illustration of the carbon cap-and-trade policy is prepared as follows

$$\delta(Z - CE) \tag{12.3}$$

where $\delta$ is the carbon tax, $Z$ is the carbon cap, and $CE$ is total carbon emissions.

### 12.5.4 Carbon Offset

The carbon offset policy is regulated to object to the carbon cap-and-trade policy (Dye and Yang 2015). The main aim of this policy is to decrease carbon emissions as much as possible (Tang et al. 2018; Halat and Hafezalkotob 2019). Following this policy, the firms are not allowed to sell their surplus allowance more emissions leads to and the higher the carbon tax imposed on the firms (Lu et al. 2020; Rout et al. 2020). The carbon emission cost function is shown as follows.

$$\delta(CE - Z) \tag{12.4}$$

where $\delta$ is the carbon tax, $Z$ is the carbon cap, and $CE$ is total carbon emissions.

### 12.5.5 Carbon Reduction Investment

Carbon reduction investment is an approach employed by the firms to decrease carbon emissions as much as possible in the long-term (Jaber et al. 2013; Toptal et al. 2014). Following this approach, some filters and sensors can be purchased by the firms to recognize the carbon emissions and decrease emissions (Lou et al. 2015; Sarkar et al. 2016). Moreover, carbon reduction investment will be added to the inventory costs (Tang et al. 2018; Cao et al. 2017). Also, a decreasing function such as $f(\xi)$ will be multiplied by the carbon emissions function to decrease the emissions (Mishra et al. 2020a, b). The carbon emission function considering reduction investment is exposed as follows

$$\delta(CE)f(\xi) \tag{12.5}$$

where $\delta$ is the carbon tax, $CE$ is total carbon emissions, and $f(\xi)$ is the carbon reduction investment function which is a decreasing function ($f^{'}(\xi) < 0$).

### 12.5.6 A Sample Mathematical Model

In this section, we propose the first mathematical model proposed by Mishra et al. (2020a) as a sample. In this paper, the annual costs related to carbon emissions are defined as follows.

Annual carbon reduction investment: $GTI = \frac{GT}{T} = G$

Total allowable carbon emissions: $CA = \delta Z$

The carbon emission associated with setup cost: $SCEC = \frac{\delta e_A}{T}$

The carbon emission associated with holding the inventory: $HCEC = \frac{\delta e_{OBS} bDT}{2}$

The carbon emission associated with disposing of the obsolete items: $OBSEC = \frac{\delta e_H a\alpha DT}{2}$

The carbon emission associated with manufacturing $CPEC = \frac{\delta e_M DT}{T} = \delta e_M D$

The carbon emission associated with remanufacturing $EICS = \frac{\delta e_R DT}{T} = \delta e_R D$.

Therefore, the total profit function under a carbon cap-and-trade policy is calculated as follows

$$\begin{aligned} TP = {} & \text{Relevant inventory costs} \\ & + \delta[Z - (SCEC + HCEC + OBSEC + CPEC + EICS)] \end{aligned} \tag{12.6}$$

## 12.6 Summary

According to the abovementioned discussions on the importance of carbon emission policies in inventory management systems, all the 75 papers studied for this research are classified according to the carbon policies they attended, the causes of carbon emissions in the inventory system they studied, the demand function type

**Table 12.1** Notations

| Notation | Description |
|---|---|
| $I$ | Maximum amount of inventory (unit/year) |
| $\delta$ | Carbon tax ($/kg) |
| $Z$ | Carbon cap (kg) |
| $e_A$ | Carbon emissions unit associated with setup cost (kg/year) |
| $e_H$ | Carbon emissions unit associated with holding the inventory (kg/year) |
| $e_{OBS}$ | Carbon emissions unit associated with disposing of the obsolete items (kg/year) |
| $e_M$ | Carbon emissions unit associated with manufacturing (kg/year) |
| $e_R$ | Carbon emissions unit associated with remanufacturing (kg/year) |
| $\alpha$ | Inventory obsolescence rate (%) |
| $a$ | Obsolete inventory weight in the warehouse (kg/unit) |
| $b$ | Space required for each unit of product (meters/unit) |

*Source* own

they considered, the existence of deteriorating or defective items, trade credit policy, and allowable shortages. The papers are summarized in Table 12.1 (Table 12.2).

## 12.7 Concluding Remarks

In today's market, environmental considerations are attending by the firms because of governmental and non-governmental pressures and consumers tend to use the items that are manufactured in sustainable ways. In general, all the regulations are making effort to eliminate carbon emissions from the inventory systems.

This paper proposed a systematic literature review that emphasized the role of carbon emission regulations on inventory management systems. By analyzing 75 papers related to this topic, a classification of different aspects of a paper such as the reasons for carbon emission, the policies regulated, and the demand function utilized is derived. The outcome of these findings is summarized in statistical analyses. In addition, different carbon emission policies are proposed and their roles and employments as a part of the mathematical models are identified.

There are few works attended carbon emissions due to setup, remanufacturing, shortage, and deterioration. Moreover, carbon cap and carbon offset policies are the ones that can be utilized in future works. Regarding the demand function type, the demand functions that are stock-dependent, credit-dependent, and promotional effort-dependent are scarce in the literature.

For future research, considering items with different deterioration rates such as items with maximum lifetime or stochastic deterioration can add value to the models. Another novelty can be using the items with defective items. Investing in reducing

**Table 12.2** Summary of papers reviewed in the study

| Paper | Carbon emission policies | | | | | Carbon emission causes | |
|---|---|---|---|---|---|---|---|
| | Carbon cap | Carbon tax (cost) | Carbon trade | Carbon offset | Carbon reduction investment | Ordering/ Replenishment | Storage/ Warehousing |
| Hua et al. (2011) | ✓ | | ✓ | | | ✓ | ✓ |
| Bouchery et al. (2012) | | ✓ | | | | ✓ | ✓ |
| Arslan and Turkay (2013) | ✓ | ✓ | ✓ | ✓ | | ✓ | ✓ |
| Chen et al. (2013) | ✓ | ✓ | | ✓ | | ✓ | ✓ |
| Jaber et al. (2013) | | ✓ | | | ✓ | | |
| Krass et al. (2013) | | ✓ | | | ✓ | | |
| Battini et al. (2014) | | ✓ | | | | | ✓ |
| Bozorgi et al. (2014) | | ✓ | | | | | ✓ |
| Konur et al. (2017) | ✓ | ✓ | | | | ✓ | ✓ |
| Toptal et al. (2014) | ✓ | ✓ | ✓ | | ✓ | ✓ | ✓ |
| Bazan et al. (2015a) | | ✓ | | | | | |
| Bazan et al. (2015b) | ✓ | ✓ | | | | | |
| Chang et al. (2015) | ✓ | | | | ✓ | | |
| Dye and Yang (2015) | ✓ | | ✓ | ✓ | | ✓ | ✓ |
| Gurtu et al. (2015) | | ✓ | | | | | |
| He et al. (2015) | ✓ | ✓ | ✓ | | | | ✓ |
| Hovelaque and Bironneau (2015) | | ✓ | | | | ✓ | ✓ |
| Lou et al. (2015) | ✓ | | ✓ | | ✓ | | |
| Qin et al. (2015) | ✓ | | ✓ | | | ✓ | ✓ |
| Schaefer and Konur (2015) | ✓ | ✓ | | | | ✓ | ✓ |
| Chen et al. (2016) | ✓ | | ✓ | | ✓ | | ✓ |
| Dong et al. (2016) | ✓ | ✓ | ✓ | | ✓ | ✓ | |

(continued)

Table 12.2 (continued)

| Paper | Carbon emission policies | | | | | Carbon emission causes | |
|---|---|---|---|---|---|---|---|
| | Carbon cap | Carbon tax (cost) | Carbon trade | Carbon offset | Carbon reduction investment | Ordering/ Replenishment | Storage/ Warehousing |
| Ji et al. (2011) | | ✓ | | | ✓ | | |
| Xu et al. (2016a) | ✓ | | ✓ | | ✓ | | |
| Xu et al. (2016b) | ✓ | ✓ | ✓ | | | | |
| Bouchery et al. (2017) | | ✓ | | | | ✓ | ✓ |
| Cao et al. (2017) | ✓ | | ✓ | | ✓ | | |
| Datta (2017) | | ✓ | | | ✓ | | ✓ |
| Ghosh et al. (2017) | ✓ | | | | | | ✓ |
| Ji et al. (2011) | | ✓ | | | ✓ | | |
| Konur et al. (2017) | ✓ | ✓ | | | | ✓ | ✓ |
| Marklund and Berling (2017) | ✓ | | ✓ | | | ✓ | ✓ |
| Qi et al. (2017) | ✓ | | | | | ✓ | ✓ |
| Sabzevar et al. (2017) | ✓ | | ✓ | | | | |
| Shu et al. (2017) | ✓ | | ✓ | | | | |
| Tsao et al. (2017) | ✓ | ✓ | ✓ | | | | ✓ |
| Wang et al. (2017) | ✓ | | ✓ | | | | |
| Xu et al. (2017) | ✓ | | ✓ | | | | |
| Yang et al. (2017) | ✓ | | ✓ | | ✓ | | |
| Aljazzar et al. (2018) | | ✓ | | | | | |
| Bai et al. (2018) | ✓ | | ✓ | | ✓ | | |
| Battini et al. (2018) | ✓ | | ✓ | | ✓ | | ✓ |
| Darom et al. (2018) | | ✓ | | | | | |

(continued)

**Table 12.2** (continued)

| Paper | Carbon emission policies | | | | | Carbon emission causes | |
|---|---|---|---|---|---|---|---|
| | Carbon cap | Carbon tax (cost) | Carbon trade | Carbon offset | Carbon reduction investment | Ordering/ Replenishment | Storage/ Warehousing |
| Huang et al. (2018) | | ✓ | | | | | ✓ |
| Kazemi et al. (2018) | | ✓ | | | | | ✓ |
| Liao and Deng (2018a) | ✓ | | ✓ | | | ✓ | ✓ |
| Liao and Deng (2018a) | | ✓ | | | | | |
| Xu et al. (2018) | | ✓ | | | | | ✓ |
| Sarkar et al. (2018) | | ✓ | | | | | ✓ |
| Sarkar et al. (2018b) | | ✓ | | | | | |
| Taleizadeh et al. (2018) | | ✓ | | | | | ✓ |
| Tang et al. (2018) | | ✓ | | ✓ | ✓ | | ✓ |
| Tiwari et al. (2018) | | ✓ | | | | | ✓ |
| Turki et al. (2018) | ✓ | | ✓ | | | | |
| Xu et al. (2018) | ✓ | | ✓ | | | ✓ | ✓ |
| Zhou et al. (2018) | | ✓ | | | | | |
| Aliabadi et al. (2019) | | ✓ | | | | ✓ | ✓ |
| Bai et al. (2019) | ✓ | | ✓ | | ✓ | | ✓ |
| Daryanto et al. (2019) | | ✓ | | | | | ✓ |
| Halat and Hafezalkotob (2019) | ✓ | ✓ | ✓ | ✓ | | ✓ | ✓ |
| Jabbarzadeh et al. (2019) | | ✓ | | | | ✓ | ✓ |
| Li and Hai (2019) | | ✓ | | | | ✓ | ✓ |
| Shi et al. (2019) | | ✓ | | | ✓ | | |
| Bai et al. (2020) | | ✓ | | | ✓ | | |

(continued)

**Table 12.2** (continued)

| Paper | Carbon emission policies | | | | | Carbon emission causes | |
|---|---|---|---|---|---|---|---|
| | Carbon cap | Carbon tax (cost) | Carbon trade | Carbon offset | Carbon reduction investment | Ordering/ Replenishment | Storage/ Warehousing |
| Datta et al. (2020) | | ✓ | | | ✓ | | ✓ |
| Heydari and Mirzajani (2020) | ✓ | | ✓ | | | | |
| Kang et al. (2020) | | | | | ✓ | | |
| Lu et al. (2020) | ✓ | | ✓ | ✓ | ✓ | ✓ | ✓ |
| Mishra et al. (2020a) | ✓ | ✓ | | | ✓ | | ✓ |
| Mishra et al. (2020b) | ✓ | | ✓ | | ✓ | | |
| Rout et al. (2020) | ✓ | ✓ | ✓ | ✓ | | | ✓ |
| Shi et al. (2020) | | ✓ | | | | | ✓ |
| Taleizadeh et al. (2020) | | ✓ | | | | ✓ | ✓ |
| Wang and Hui (2020) | | | | | ✓ | | |
| Wangsa et al. (2020) | | ✓ | | | | | ✓ |
| Sepehri (2021) | ✓ | ✓ | | | | ✓ | ✓ |
| Sepehri et al. (2021) | ✓ | ✓ | | | ✓ | ✓ | ✓ |

| Paper | Carbon emission causes | | | | | | |
|---|---|---|---|---|---|---|---|
| | Purchasing/ Procurement | Setup | Transportation | Manufacturing | Remanufacturing/ Recycling | Disposing of the obsolete items | Shortage |
| Hua et al. (2011) | | | | | | | |
| Bouchery et al. (2012) | | | | | | | |
| Arslan and Turkay (2013) | | | | | | | |
| Chen et al. (2013) | ✓ | | | | | | |
| Jaber et al. (2013) | | | | ✓ | | | |

(continued)

**Table 12.2** (continued)

| Paper | Carbon emission causes | | | | | | |
|---|---|---|---|---|---|---|---|
| | Purchasing/ Procurement | Setup | Transportation | Manufacturing | Remanufacturing/ Recycling | Disposing of the obsolete items | Shortage |
| Krass et al. (2013) | | | | ✓ | | | |
| Battini et al. (2014) | | | ✓ | | | ✓ | |
| Bozorgi et al. (2014) | | | ✓ | | | | |
| Konur et al. (2017) | ✓ | | ✓ | | | | |
| Toptal et al. (2014) | ✓ | | | | | | |
| Bazan et al. (2015a) | | | ✓ | ✓ | ✓ | | |
| Bazan et al. (2015b) | | | ✓ | ✓ | | | |
| Chang et al. (2015) | | | | ✓ | ✓ | | |
| Dye and Yang (2015) | ✓ | | | | | | |
| Gurtu et al. (2015) | | | ✓ | | | | |
| He et al. (2015) | | ✓ | | ✓ | | | |
| Hovelaque and Bironneau (2015) | | | | | | | |
| Lou et al. (2015) | | | | ✓ | | | |
| Qin et al. (2015) | ✓ | | | | | | |
| Schaefer and Konur (2015) | ✓ | | | | | | ✓ |
| Chen et al. (2016) | | | | | | | |
| Dong et al. (2016) | | | | | | | |
| Sarkar et al. () | | | ✓ | | | | |
| Xu et al. (2016a) | | | | ✓ | | | |
| Xu et al. (2016b) | | | | ✓ | | | |
| Bouchery et al. (2017) | | | | | | | |

(continued)

**Table 12.2** (continued)

| Paper | Carbon emission causes | | | | | | |
|---|---|---|---|---|---|---|---|
| | Purchasing/ Procurement | Setup | Transportation | Manufacturing | Remanufacturing/ Recycling | Disposing of the obsolete items | Shortage |
| Cao et al. (2017) | | | | ✓ | | | |
| Datta (2017) | | | | ✓ | | ✓ | |
| Ghosh et al. (2017) | | ✓ | ✓ | ✓ | | | |
| Ji et al. (2011) | | | | ✓ | | | |
| Konur et al. (2017) | ✓ | | ✓ | | | | ✓ |
| Marklund and Berling (2017) | ✓ | | | | | | |
| Qi et al. (2017) | | | | | | | |
| Sabzevar et al. (2017) | | | | | | | |
| Shu et al. (2017) | | | ✓ | ✓ | ✓ | | |
| Tsao et al. (2017) | | | ✓ | | ✓ | ✓ | |
| Wang et al. (2017) | | | | ✓ | ✓ | | |
| Xu et al. (2017) | | | | ✓ | | | |
| Yang et al. (2017) | | | | ✓ | | | |
| Aljazzar et al. (2018) | | | ✓ | ✓ | | | |
| Bai et al. (2018) | | | | ✓ | | | |
| Battini et al. (2018) | | | ✓ | | | ✓ | |
| Darom et al. (2018) | | | ✓ | | | | |
| Huang et al. (2018) | | | ✓ | ✓ | | | |
| Kazemi et al. (2018) | | | | | | ✓ | |
| Liao and Deng (2018a) | | | | | | | |
| Liao and Deng (2018a) | | | | ✓ | ✓ | | |

(continued)

**Table 12.2** (continued)

| Paper | Carbon emission causes | | | | | | |
|---|---|---|---|---|---|---|---|
| | Purchasing/ Procurement | Setup | Transportation | Manufacturing | Remanufacturing/ Recycling | Disposing of the obsolete items | Shortage |
| Xu et al. (2018) | | | ✓ | | | ✓ | |
| Sarkar et al. (2018) | | | | | | | |
| Sarkar et al. (2018b) | | | ✓ | | | | |
| Taleizadeh et al. (2018) | | | | ✓ | | ✓ | |
| Tang et al. (2018) | | | ✓ | | | | |
| Tiwari et al. (2018) | | | ✓ | | | ✓ | |
| Turki et al. (2018) | | | | ✓ | ✓ | | |
| Xu et al. (2018) | | | | | | | |
| Zhou et al. (2018) | | | | | ✓ | | |
| Aliabadi et al. (2019) | | | | | | | |
| Bai et al. (2019) | | | | ✓ | | | |
| Daryanto et al. (2019) | | | ✓ | | | ✓ | |
| Halat and Hafezalkotob (2019) | | | ✓ | ✓ | | | |
| Jabbarzadeh et al. (2019) | | | | | | | |
| Li and Hai (2019) | | | | | | | |
| Shi et al. (2019) | | | | ✓ | | | |
| Bai et al. (2020) | | | | ✓ | | | |
| Datta et al. (2020) | | ✓ | | ✓ | | | |
| Heydari and Mirzajani (2020) | | | | ✓ | | | |
| Kang et al. (2020) | | | | ✓ | | | |
| Lu et al. (2020) | ✓ | ✓ | ✓ | ✓ | | | |

(continued)

**Table 12.2** (continued)

| Paper | Carbon emission causes | | | | | | |
|---|---|---|---|---|---|---|---|
| | Purchasing/ Procurement | Setup | Transportation | Manufacturing | Remanufacturing/ Recycling | Disposing of the obsolete items | Shortage |
| Mishra et al. (2020a) | | ✓ | | ✓ | ✓ | ✓ | |
| Mishra et al. (2020b) | ✓ | | | | | | |
| Rout et al. (2020) | | ✓ | ✓ | ✓ | ✓ | ✓ | |
| Shi et al. (2020) | ✓ | | ✓ | ✓ | | | |
| Taleizadeh et al. (2020) | | | ✓ | | | ✓ | |
| Wang and Hui (2020) | | | | ✓ | | | |
| Wangsa et al. (2020) | | | ✓ | ✓ | | | |
| Sepehri (2021) | | | | | | | |
| Sepehri et al. (2021) | | | | | | | |

| Paper | Carbon emission causes | Demand | | | | | |
|---|---|---|---|---|---|---|---|
| | Deterioration | Constant rate | Price-dependent | Credit-dependent | Stock-dependent | Promotional effort-dependent | Carbon emission-dependent |
| Hua et al. (2011) | | ✓ | | | | | |
| Bouchery et al. (2012) | | ✓ | | | | | |
| Arslan and Turkay (2013) | | ✓ | | | | | |
| Chen et al. (2013) | | ✓ | | | | | |
| Jaber et al. (2013) | | ✓ | | | | | |
| Krass et al. (2013) | | | ✓ | | | | |
| Battini et al. (2014) | | ✓ | | | | | |
| Bozorgi et al. (2014) | | ✓ | | | | | |

(continued)

**Table 12.2** (continued)

| Paper | Carbon emission causes | Demand | | | | | |
|---|---|---|---|---|---|---|---|
| | Deterioration | Constant rate | Price-dependent | Credit-dependent | Stock-dependent | Promotional effort-dependent | Carbon emission-dependent |
| Konur et al. (2017) | | ✓ | | | | | |
| Toptal et al. (2014) | | ✓ | | | | | |
| Bazan et al. (2015a) | | ✓ | | | | | |
| Bazan et al. (2015b) | | ✓ | | | | | |
| Chang et al. (2015) | | ✓ | | | | | |
| Dye and Yang (2015) | | | | ✓ | | | |
| Gurtu et al. (2015) | | ✓ | | | | | |
| He et al. (2015) | | ✓ | | | | | |
| Hovelaque and Bironneau (2015) | | | ✓ | | | | ✓ |
| Lou et al. (2015) | | | ✓ | | | | ✓ |
| Qin et al. (2015) | | | | ✓ | | | ✓ |
| Schaefer and Konur (2015) | | | | | | | |
| Chen et al. (2016) | | | | | | | |
| Dong et al. (2016) | | | | | | | ✓ |
| Sarkar et al. () | | ✓ | | | | | |
| Xu et al. (2016a) | | | ✓ | | | | ✓ |
| Xu et al. (2016b) | | | ✓ | | | | |
| Bouchery et al. (2017) | | ✓ | | | | | |
| Cao et al. (2017) | | | ✓ | | | | ✓ |
| Datta (2017) | | | ✓ | | | | |
| Ghosh et al. (2017) | | | | | | | |

(continued)

**Table 12.2** (continued)

| Paper | Carbon emission causes | Demand | | | | | |
|---|---|---|---|---|---|---|---|
| | Deterioration | Constant rate | Price-dependent | Credit-dependent | Stock-dependent | Promotional effort-dependent | Carbon emission-dependent |
| Ji et al. (2011) | | | ✓ | | | | ✓ |
| Konur et al. (2017) | | | | | | | |
| Marklund and Berling (2017) | | ✓ | | | | | |
| Qi et al. (2017) | | | ✓ | | | | |
| Sabzevar et al. (2017) | | | ✓ | | | | |
| Shu et al. (2017) | | ✓ | | | | | |
| Tsao et al. (2017) | | | ✓ | ✓ | | | |
| Wang et al. (2017) | | | | | | | |
| Xu et al. (2017) | | | ✓ | | | | |
| Yang et al. (2017) | | | ✓ | | | | ✓ |
| Aljazzar et al. (2018) | | ✓ | | | | | |
| Bai et al. (2018) | | | ✓ | | | | |
| Battini et al. (2018) | | ✓ | | | | | |
| Darom et al. (2018) | | ✓ | | | | | |
| Huang et al. (2018) | ✓ | | ✓ | | | | |
| Kazemi et al. (2018) | | ✓ | | | | | |
| Liao and Deng (2018a) | | | | | | | |
| Liao and Deng (2018a) | | | | | | | |
| Xu et al. (2018) | | ✓ | | | | | |
| Sarkar et al. (2018) | | ✓ | | | | | |
| Sarkar et al. (2018b) | | ✓ | | | | | |

(continued)

**Table 12.2** (continued)

| Paper | Carbon emission causes | Demand | | | | | |
|---|---|---|---|---|---|---|---|
| | Deterioration | Constant rate | Price-dependent | Credit-dependent | Stock-dependent | Promotional effort-dependent | Carbon emission-dependent |
| Taleizadeh et al. (2018) | | ✓ | | | | | |
| Tang et al. (2018) | | | | | | | |
| Tiwari et al. (2018) | | ✓ | | | | | |
| Turki et al. (2018) | | | | | | | |
| Xu et al. (2018) | | | | | | | |
| Zhou et al. (2018) | | | | | | | |
| Aliabadi et al. (2019) | | | ✓ | ✓ | | | ✓ |
| Bai et al. (2019) | | | ✓ | | | | |
| Daryanto et al. (2019) | | ✓ | | | | | |
| Halat and Hafezalkotob (2019) | | ✓ | | | | | |
| Jabbarzadeh et al. (2019) | | | ✓ | | | ✓ | |
| Li and Hai (2019) | | ✓ | | | | | |
| Shi et al. (2019) | | | ✓ | | | | |
| Bai et al. (2020) | | | ✓ | | | ✓ | ✓ |
| Datta et al. (2020) | | | ✓ | | | | |
| Heydari and Mirzajani (2020) | | | | | | | |
| Kang et al. (2020) | | | ✓ | | | | ✓ |
| Lu et al. (2020) | | ✓ | | | | | |
| Mishra et al. (2020a) | | ✓ | | | | | |
| Mishra et al. (2020b) | | | ✓ | | ✓ | | |
| Rout et al. (2020) | ✓ | ✓ | | | | | |

(continued)

**Table 12.2** (continued)

| Paper | Carbon emission causes | Demand | | | | | |
|---|---|---|---|---|---|---|---|
| | Deterioration | Constant rate | Price-dependent | Credit-dependent | Stock-dependent | Promotional effort-dependent | Carbon emission-dependent |
| Shi et al. (2020) | | ✓ | | | | | |
| Taleizadeh et al. (2020) | | | ✓ | | | | |
| Wang and Hui (2020) | | | ✓ | | | | ✓ |
| Wangsa et al. (2020) | | | | | | | |
| Sepehri (2021) | | | ✓ | | | | |
| Sepehri et al. (2021) | | | ✓ | | | | |

| Paper | Demand | Deteriorating items | Trade credit policy | Shortage | Defective (Imperfect quality) items |
|---|---|---|---|---|---|
| | Stochastic | | | | |
| Hua et al. (2011) | | | | | |
| Bouchery et al. (2012) | | | | | |
| Arslan and Turkay (2013) | | | | | |
| Chen et al. (2013) | | | | | |
| Jaber et al. (2013) | | | | | |
| Krass et al. (2013) | | | | | |
| Battini et al. (2014) | | | | | |
| Bozorgi et al. (2014) | | | | | |
| Konur et al. (2017) | | | | | |
| Toptal et al. (2014) | | | | | |
| Bazan et al. (2015a) | | | | | |
| Bazan et al. (2015b) | | | | | |
| Chang et al. (2015) | | | | | |

(continued)

**Table 12.2** (continued)

| Paper | Demand | Deteriorating items | Trade credit policy | Shortage | Defective (Imperfect quality) items |
|---|---|---|---|---|---|
| | Stochastic | | | | |
| Dye and Yang (2015) | | ✓ | ✓ | ✓ | |
| Gurtu et al. (2015) | | | | | |
| He et al. (2015) | | | | | |
| Hovelaque and Bironneau (2015) | | | | | |
| Lou et al. (2015) | | | | | |
| Qin et al. (2015) | | | ✓ | | |
| Schaefer and Konur (2015) | ✓ | | | ✓ | |
| Chen et al. (2016) | ✓ | | | | |
| Dong et al. (2016) | | | | | |
| Sarkar et al. () | | | | | ✓ |
| Xu et al. (2016a) | | | | | |
| Xu et al. (2016b) | | | | | |
| Bouchery et al. (2017) | | | | | |
| Cao et al. (2017) | | | | | |
| Datta (2017) | | | | | ✓ |
| Ghosh et al. (2017) | ✓ | | | ✓ | |
| Ji et al. (2011) | | | | | |
| Konur et al. (2017) | ✓ | | ✓ | | |
| Marklund and Berling (2017) | ✓ | | | | |
| Qi et al. (2017) | | | | | |
| Sabzevar et al. (2017) | | | | | |

(continued)

Table 12.2 (continued)

| Paper | Demand | Deteriorating items | Trade credit policy | Shortage | Defective (Imperfect quality) items |
|---|---|---|---|---|---|
| | Stochastic | | | | |
| Shu et al. (2017) | | | | | ✓ |
| Tsao et al. (2017) | | | ✓ | | |
| Wang et al. (2017) | ✓ | | | | |
| Xu et al. (2017) | | | | | |
| Yang et al. (2017) | | | | | |
| Aljazzar et al. (2018) | | | ✓ | | |
| Bai et al. (2018) | | | | | |
| Battini et al. (2018) | | | | | |
| Darom et al. (2018) | | | | ✓ | |
| Huang et al. (2018) | | ✓ | | | |
| Kazemi et al. (2018) | | | | | ✓ |
| Liao and Deng (2018a) | ✓ | | | | |
| Liao and Deng (2018a) | ✓ | | | | |
| Xu et al. (2018) | | | | ✓ | |
| Sarkar et al. (2018) | | | | ✓ | ✓ |
| Sarkar et al. (2018b) | | | ✓ | | |
| Taleizadeh et al. (2018) | | | | ✓ | |
| Tang et al. (2018) | ✓ | | | ✓ | |
| Tiwari et al. (2018) | | ✓ | | | ✓ |
| Turki et al. (2018) | ✓ | | | ✓ | |
| Xu et al. (2018) | ✓ | | | | |

(continued)

**Table 12.2** (continued)

| Paper | Demand | Deteriorating items | Trade credit policy | Shortage | Defective (Imperfect quality) items |
|---|---|---|---|---|---|
| | Stochastic | | | | |
| Zhou et al. (2018) | ✓ | | | | |
| Aliabadi et al. (2019) | | ✓ | ✓ | ✓ | |
| Bai et al. (2019) | | ✓ | | | |
| Daryanto et al. (2019) | | ✓ | | | |
| Halat and Hafezalkotob (2019) | | | | | |
| Jabbarzadeh et al. (2019) | | | ✓ | ✓ | |
| Li and Hai (2019) | | | | | |
| Shi et al. (2019) | | | | | |
| Bai et al. (2020) | ✓ | | | | |
| Datta et al. (2020) | | | | | |
| Heydari and Mirzajani (2020) | ✓ | | | ✓ | |
| Kang et al. (2020) | | | | | |
| Lu et al. (2020) | | ✓ | | | |
| Mishra et al. (2020a) | | | | ✓ | |
| Mishra et al. (2020b) | | ✓ | | | |
| Rout et al. (2020) | | ✓ | | | ✓ |
| Shi et al. (2020) | | ✓ | ✓ | | |
| Taleizadeh et al. (2020) | | | | ✓ | |
| Wang and Hui (2020) | | | | | |
| Wangsa et al. (2020) | ✓ | | | ✓ | |
| Sepehri (2021) | | ✓ | ✓ | | |
| Sepehri et al. (2021) | | ✓ | ✓ | | |

*Source* own

different relevant costs such as setup cost is another new value that can be added to the future works.

## References

Aliabadi L, Yazdanparast R, Nasiri MM (2019) An inventory model for non-instantaneous deteriorating items with credit period and carbon emission sensitive demand: a signomial geometric programming approach. Int J Manage Sci Eng Manage 14:124–136
Aljazzar SM, Gurtu A, Jaber MY (2018) Delay-in-payments—a strategy to reduce carbon emissions from supply chains. J Clean Prod 170:636–644
Arslan MC, Turkay M (2013) EOQ revisited with sustainability considerations. Found Comput Decis Sci 38:223–249
Bai Q, Xu J, Zhang Y (2018) Emission reduction decision and coordination of a make-to-order supply chain with two products under cap-and-trade regulation. Comput Ind Eng 119:131–145
Bai Q, Jin M, Xu X (2019) Effects of carbon emission reduction on supply chain coordination with vendor-managed deteriorating product inventory. Int J Prod Econ 208:83–99
Bai Q, Xu J, Chauhan SS (2020) Effects of sustainability investment and risk aversion on a two-stage supply chain coordination under a carbon tax policy. Comput Industr Eng 142:106324
Bansal JC, Das KN, Nagar A, Deep K, Ojha AK (2018) Soft computing for problem solving: SocProS 2017, vols 1, 816. Springer
Battini D, Persona A, Sgarbossa F (2014) A sustainable EOQ model: theoretical formulation and applications. Int J Prod Econ 149:145–153
Battini D, Calzavara M, Isolan I, Sgarbossa F, Zangaro F (2018) Sustainability in material purchasing: a multi-objective economic order quantity model under carbon trading. Sustainability 10:4438
Bazan E, Jaber MY, El Saadany AM (2015a) Carbon emissions and energy effects on manufacturing–remanufacturing inventory models. Comput Industr Eng 88:307–316
Bazan E, Jaber MY, Zanoni S (2015b) Supply chain models with greenhouse gases emissions, energy usage and different coordination decisions. Appl Math Model 39:5131–5151
Benjaafar S, Li Y, Daskin M (2012) Carbon footprint and the management of supply chains: insights from simple models. IEEE Trans Autom Sci Eng 10:99–116
Bonney M, Jaber MY (2011) Environmentally responsible inventory models: non-classical models for a non-classical era. Int J Prod Econ 133:43–53
Bouchery Y, Ghaffari A, Jemai Z, Dallery Y (2012) Including sustainability criteria into inventory models. Eur J Oper Res 222:229–240
Bouchery Y, Ghaffari A, Jemai Z, Tan T (2017) Impact of coordination on costs and carbon emissions for a two-echelon serial economic order quantity problem. Eur J Oper Res 260:520–533
Bozorgi A, Pazour J, Nazzal D (2014) A new inventory model for cold items that considers costs and emissions. Int J Prod Econ 155:114–125
Cao K, Xu X, Wu Q, Zhang Q (2017) Optimal production and carbon emission reduction level under cap-and-trade and low carbon subsidy policies. J Clean Prod 167:505–513
Chang X, Xia H, Zhu H, Fan T, Zhao H (2015) Production decisions in a hybrid manufacturing–remanufacturing system with carbon cap and trade mechanism. Int J Prod Econ 162:160–173
Chen X, Benjaafar S, Elomri A (2013) The carbon-constrained EOQ. Oper Res Lett 41:172–179
Chen X, Wang X, Kumar V, Kumar N (2016) Low carbon warehouse management under cap-and-trade policy. J Clean Prod 139:894–904
Darom NA, Hishamuddin H, Ramli R, Nopiah ZM (2018) An inventory model of supply chain disruption recovery with safety stock and carbon emission consideration. J Clean Prod 197:1011–1021

Daryanto Y, Wee HM, Astanti RD (2019) Three-echelon supply chain model considering carbon emission and item deterioration. Transp Res Part E Logist Transp Rev 122:368–383
Datta TK (2017) Effect of green technology investment on a production-inventory system with carbon tax. Adv Oper Res
Datta TK, Nath P, Choudhury KD (2020) A hybrid carbon policy inventory model with emission source-based green investments. OPSEARCH 57:202–220
Dekker R, Bloemhof J, Mallidis I (2012) Operations research for green logistics—an overview of aspects, issues, contributions and challenges. Euro J Oper Res 219:671–679
Dong C, Shen B, Chow P-S, Yang L, Ng CT (2016) Sustainability investment under cap-and-trade regulation. Ann Oper Res 240:509–531
Dye C-Y, Yang C-T (2015) Sustainable trade credit and replenishment decisions with credit-linked demand under carbon emission constraints. Eur J Oper Res 244:187–200
Ghosh A, Jha J, Sarmah S (2017) Optimal lot-sizing under strict carbon cap policy considering stochastic demand. Appl Math Model 44:688–704
Gupta MC (1995) Environmental management and its impact on the operations function. Int J Oper Prod Manage
Gurtu A, Jaber MY, Searcy C (2015) Impact of fuel price and emissions on inventory policies. Appl Math Model 39:1202–1216
Halat K, Hafezalkotob A (2019) Modeling carbon regulation policies in inventory decisions of a multi-stage green supply chain: a game theory approach. Comput Ind Eng 128:807–830
He P, Zhang W, Xu X, Bian Y (2015) Production lot-sizing and carbon emissions under cap-and-trade and carbon tax regulations. J Clean Prod 103:241–248
Heydari J, Mirzajani Z (2020) Supply chain coordination under nonlinear cap and trade carbon emission function and demand uncertainty. Kybernetes
Hovelaque V, Bironneau L (2015) The carbon-constrained EOQ model with carbon emission dependent demand. Int J Prod Econ 164:285–291
Hua G, Cheng T, Wang S (2011) Managing carbon footprints in inventory management. Int J Prod Econ 132:178–185
Huang H, He Y, Li D (2018) Pricing and inventory decisions in the food supply chain with production disruption and controllable deterioration. J Clean Prod 180:280–296
Iqbal MW, Sarkar B (2019) Recycling of lifetime dependent deteriorated products through different supply chains. RAIRO Oper Res 53(1):129–156
Iqbal MW, Sarkar B (2020) Application of preservation technology for lifetime dependent products in an integrated production system. J Industr Manage Optimization 16(1):141
Jabbarzadeh A, Aliabadi L, Yazdanparast R (2019) Optimal payment time and replenishment decisions for retailer's inventory system under trade credit and carbon emission constraints. Oper Res 1–32
Jaber MY, Glock CH, El Saadany AM (2013) Supply chain coordination with emissions reduction incentives. Int J Prod Res 51:69–82
Ji J, Zhang Z, Yang L (2017) Carbon emission reduction decisions in the retail-/dual-channel supply chain with consumers' preference. J Clean Prod 141:852–867
Jia F, Zhang T, Chen L (2020) Sustainable supply chain finance: towards a research agenda. J Clean Prod 243:118680
Kamble SS, Gunasekaran A, Gawankar SA (2018) Sustainable industry 4.0 framework: a systematic literature review identifying the current trends and future perspectives. Process Saf Environ Protection 117:408–425
Kang H, Jung S-Y, Lee H (2020) The impact of green credit policy on manufacturers' efforts to reduce suppliers' pollution. J Clean Prod 248:119271
Kazemi N, Abdul-Rashid SH, Ghazilla RAR, Shekarian E, Zanoni S (2018) Economic order quantity models for items with imperfect quality and emission considerations. Int J Syst Sci Oper Logist 5:99–115
Kleindorfer PR, Singhal K, Van Wassenhove LN (2005) Sustainable operations management. Prod Oper Manage 14:482–492

Konur D (2014) Carbon constrained integrated inventory control and truckload transportation with heterogeneous freight trucks. Int J Prod Econ 153:268–279
Konur D, Campbell JF, Monfared SA (2017) Economic and environmental considerations in a stochastic inventory control model with order splitting under different delivery schedules among suppliers. Omega 71:46–65
Krass D, Nedorezov T, Ovchinnikov A (2013) Environmental taxes and the choice of green technology. Prod Oper Manage 22:1035–1055
Li Z, Hai J (2019) Inventory management for one warehouse multi-retailer systems with carbon emission costs. Comput Ind Eng 130:565–574
Liao H, Deng Q (2018a) A carbon-constrained EOQ model with uncertain demand for remanufactured products. J Clean Prod 199:334–347
Liao H, Deng Q (2018b) EES-EOQ model with uncertain acquisition quantity and market demand in dedicated or combined remanufacturing systems. Appl Math Model 64:135–167
Lin H-J (2018) Investing in transportation emission cost reduction on environmentally sustainable EOQ models with partial backordering. J Appl Sci Eng 21:291–303
Linton JD, Klassen R, Jayaraman V (2007) Sustainable supply chains: an introduction. J Oper Manage 25:1075–1082
Lou GX, Xia HY, Zhang JQ, Fan TJ (2015) Investment strategy of emission-reduction technology in a supply chain. Sustainability 7:10684–10708
Lu C-J, Yang C-T, Yen H-F (2020) Stackelberg game approach for sustainable production-inventory model with collaborative investment in technology for reducing carbon emissions. J Clean Prod 121963
Marklund J, Berling P (2017) Green inventory management. In: Sustainable supply chains. Springer, pp 189–218
Mishra U, Wu J-Z, Sarkar B (2020a) A sustainable production-inventory model for a controllable carbon emissions rate under shortages. J Cleaner Production 256:120268
Mishra U, Wu J-Z, Tsao Y-C, Tseng M-L (2020b) Sustainable inventory system with controllable non-instantaneous deterioration and environmental emission rates. J Clean Prod 244:118807
Nguyen T, Li Z, Spiegler V, Ieromonachou P, Lin Y (2018) Big data analytics in supply chain management: a state-of-the-art literature review. Comput Oper Res 98:254–264
Oberthür S, Ott HE (1999) The Kyoto protocol: international climate policy for the 21st century. Springer Science & Business Media
Panayotou T (1993) Green markets: the economics of sustainable development
Pope J, Annandale D, Morrison-Saunders A (2004) Conceptualising sustainability assessment. Environ Impact Assess Rev 24:595–616
Qi Q, Wang J, Bai Q (2017) Pricing decision of a two-echelon supply chain with one supplier and two retailers under a carbon cap regulation. J Clean Prod 151:286–302
Qin J, Bai X, Xia L (2015) Sustainable trade credit and replenishment policies under the cap-and-trade and carbon tax regulations. Sustainability 7:1634–16361
Rout C, Paul A, Kumar RS, Chakraborty D, Goswami A (2020) Cooperative sustainable supply chain for deteriorating item and imperfect production under different carbon emission regulations. J Clean Prod 122170
Sabzevar N, Enns S, Bergerson J, Kettunen J (2017) Modeling competitive firms' performance under price-sensitive demand and cap-and-trade emissions constraints. Int J Prod Econ 184:193–209
Sarkar B, Saren S, Sarkar M, Seo YW (2016) A Stackelberg game approach in an integrated inventory model with carbon-emission and setup cost reduction. Sustainability 8:1244
Sarkar B, Ahmed W, Choi S-B, Tayyab M (2018a) Sustainable inventory management for environmental impact through partial backordering and multi-trade-credit-period. Sustainability 10:4761
Sarkar B, Ahmed W, Kim N (2018b) Joint effects of variable carbon emission cost and multi-delay-in-payments under single-setup-multiple-delivery policy in a global sustainable supply chain Journal of Cleaner Production 185:421–445

Schaefer B, Konur D (2015) Economic and environmental considerations in a continuous review inventory control system with integrated transportation decisions. Transp Res Part E Logist Transp Rev 80:142–165
Sepehri A (2021) Optimizing the replenishment cycle and selling price for an inventory model under carbon emission regulation and partially permissible delay in payment. Process Integr Opt Sustain 1–21
Sepehri A, Mishra U, Tseng M-L, Sarkar B (2021) Joint pricing and inventory model for deteriorating items with maximum lifetime and controllable carbon emissions under permissible delay in payments. Mathematics 9(5):470
Seuring S, Müller M (2008) From a literature review to a conceptual framework for sustainable supply chain management. J Clean Prod16:1699–1710
Shi X, Dong C, Zhang C, Zhang X (2019) Who should invest in clean technologies in a supply chain with competition? J Clean Prod 215:689–700
Shi Y, Zhang Z, Chen S-C, Cárdenas-Barrón LE, Skouri K (2020) Optimal replenishment decisions for perishable products under cash, advance, and credit payments considering carbon tax regulations. Int J Prod Econ 223:107514
Shu T, Wu Q, Chen S, Wang S, Lai KK, Yang H (2017) Manufacturers'/remanufacturers' inventory control strategies with cap-and-trade regulation. J Clean Prod 159:11–25
Srivastava SK (2007) Green supply-chain management: a state-of-the-art literature review. Int J Manag Rev 9:53–80
Taleizadeh AA, Soleymanfar VR, Govindan K (2018) Sustainable economic production quantity models for inventory systems with shortage. J Clean Prod 174:1011–1020
Taleizadeh AA, Hazarkhani B, Moon I (2020) Joint pricing and inventory decisions with carbon emission considerations, partial backordering and planned discounts. Annals Oper Res 290:95–113
Tang S, Wang W, Cho S, Yan H (2018) Reducing emissions in transportation and inventory management: (R, Q) policy with considerations of carbon reduction. Eur J Oper Res 269:327–340
Tiwari S, Daryanto Y, Wee HM (2018) Sustainable inventory management with deteriorating and imperfect quality items considering carbon emission. J Clean Prod 192:281–292
Toptal A, Özlü H, Konur D (2014) Joint decisions on inventory replenishment and emission reduction investment under different emission regulations. Int J Prod Res 52:243–269
Tranfield D, Denyer D, Smart P (2003) Towards a methodology for developing evidence-informed management knowledge by means of systematic review. Brit J Manage 14:207–222
Tsao Y-C, Lee P-L, Chen C-H, Liao Z-W (2017) Sustainable newsvendor models under trade credit. J Clean Prod 141:1478–1491
Turki S, Sauvey C, Rezg N (2018) Modelling and optimization of a manufacturing/remanufacturing system with storage facility under carbon cap and trade policy. J Clean Prod 93:441–458
Walker H, Seuring S, Sarkis J, Klassen R (2014) Sustainable operations management: recent trends and future directions. Int J Oper Prod Manage
Wang L, Hui M (2020) Research on joint emission reduction in supply chain based on carbon footprint of the product. J Clean Prod 121086
Wang Y, Chen W, Liu B (2017) Manufacturing/remanufacturing decisions for a capital-constrained manufacturer considering carbon emission cap and trade. J Clean Prod 140:1118–1128
Wangsa ID, Tiwari S, Wee HM, Reong S (2020) A sustainable vendor-buyer inventory system considering transportation, loading and unloading activities. J Clean Prod 122120
Wee H-M, Daryanto Y (2020) Imperfect quality item inventory models considering carbon emissions. In: Optimization and inventory management. Springer, pp 137–159
Xu J, Chen Y, Bai Q (2016a) A two-echelon sustainable supply chain coordination under cap-and-trade regulation. J Clean Prod 135:42–56
Xu X, Xu X, He P (2016b) Joint production and pricing decisions for multiple products with cap-and-trade and carbon tax regulations. J Clean Prod 112:4093–4106
Xu X, Zhang W, He P, Xu X (2017) Production and pricing problems in make-to-order supply chain with cap-and-trade regulation. Omega 66:248–257

Xu J, Bai Q, Xu L, Hu T (2018) Effects of emission reduction and partial demand information on operational decisions of a newsvendor problem. J Clean Prod 188:825–839

Yang L, Zhang Q, Ji J (2017) Pricing and carbon emission reduction decisions in supply chains with vertical and horizontal cooperation. Int J Prod Econ 191:286–297

Zhou J, Deng Q, Li T (2018) Optimal acquisition and remanufacturing policies considering the effect of quality uncertainty on carbon emissions. J Clean Prod 186:180–190

# Chapter 13
# Application of Triangular Fuzzy Numbers in Taking Optimal Decision

**M. Kuber Singh**

**Abstract** We shall discuss the application of triangular fuzzy numbers in the main classes of decision-making problems. In order to measure the attribute importance when individuals make a decision or evaluate the alternatives, the weight is one of the most useful tools. The problem is to select the best system from a set of finite systems $X_1, X_2, X_3, \ldots X_n$ associated with a set of objectives $G_1, G_2, G_3, \ldots G_m$ having varying degrees of importance. The selector gives his decision as to how well the alternatives satisfy the objectives in linguistic terms such as fair, good, very good, poor, etc. The objectives have different priorities given in fuzzy terms such as important, very important, less important, etc. and an aggregation is then performed using triangular fuzzy numbers. The final decision is the intersection of the given goals according to Zadeh and Bellman (Manage Sci 17:140–164, 1970). The model is based on the use of triangular fuzzy numbers. This model is applied to transport networks having varying degrees of importance available in the study area on bus priority and non-bus priority. This paper explores use of triangular fuzzy numbers in taking optimal decision.

**Keywords** Fuzzy set · Linguistic terms in fuzzy environment · Fuzzy numbers · Triangular Fuzzy numbers

**AMS Classification (2010)** 03E72

## 13.1 Introduction

Uncertainty may arise due to unclear information about the problem, or which is ambiguity. The fuzzy numbers and fuzzy values are widely used in many applications for representing uncertain information. The uncertainty arising due to vagueness can be studied by using fuzzy set theory. In 1965, Lotfi A-Zadeh propounded the concept of fuzzy set theory. The selection of best system from a set of finite number of systems

M. K. Singh (✉)
Department of Mathematics, D.M. College of Science (Dhanamanjuri University), Imphal 795001, India



N. H. Shah et al. (eds.), *Decision Making in Inventory Management*, Inventory Optimization, https://doi.org/10.1007/978-981-16-1729-4_13

had been performed by using the concept of fuzzy set. Zadeh (1965) introduced the idea of fuzzy numbers as being convex and normal fuzzy set. Zadeh's extension principle (Zadeh and Bellman 1970; Zadeh, 1978) is used for mathematical operation on real numbers and can be extended to the ones defined on fuzzy numbers. There are many researchers who have made important contribution to theory of fuzzy numbers and its applications. Some of them are Dubois et al. (1978, 1980), Kaufmann (1975), Kaufmann and Gupta (1985), and Nguyen (1978).

The optimal selection is the one that satisfies multiple goals having different objectives given in fuzzy terms. The aggregation of the priorities of these objectives is then conducted using fuzzy numbers. Zimmermann (2001) discussed the concept of the fuzzy set theory and its applications. Many applications of fuzzy set theory can be found in Zimmerman (2001). Fuzzy numbers are mostly applied to data analysis and decision-making. In particular, triangular fuzzy numbers are commonly used in applications and it is also easy to handle the difficulty. Kumar and Karmaker (2003) proposed optimal system selection using Fuzzy numbers. The concept of triangular intuitionistic fuzzy numbers (TIFN) is introduced by Feng et al. (2010). Kuchta (2010) proposed the use of fuzzy numbers in practical project planning and control. Gil-Lafuente and Merigó (2010) proposed decision-making techniques in political management. A new method for comparing fuzzy numbers base fuzzy probabilistic preference is introduced by Zhang et al. (2014). Prangishvili et al. (2015) introduced application of fuzzy sets in solving some problems of management. Brindhavanam and Rosario (2017) considered a fuzzy comparison method for a deterministic inventory model.

## 13.2 Urban Transport

Transport system currently available has shown us a lot of problems with respect to the increasing traffic. Accidents and other problems with regard to road transport are increasing day by day. Until and unless we bring forward a solution to it, it is going to be difficult for us to travel efficiently in the near future. As number of vehicles have increased, there is a lot of congestion in the streets nowadays. Accident rates are increasing day by day, and then there is also the problem of parking spaces. Hence we need to give more attention to how highway transportation is being operated and model a solution for better geometric design, capacity, traffic regulation, parking facilities, etc. Speed is a major characteristic of traffic and its measurement is important for our study. Speed is the rate of movement of traffic and is expressed in metric units in kilometers per hour. We also need to observe the flow, which is defined as the volume of traffic using the road in a given interval of time and it is expressed in vehicles per hour or vehicles per day. By understanding the flow characteristics we can determine if a particular road is handling traffic at its capacity or not. If the traffic is heavy, it leads to traffic congestion which results in lower travel speed. Lower speeds in turn lead to economic loss because of time lost by the people traveling by the vehicles and higher operational cost of vehicles. Traffic congestions also lead to environmental pollution. Volume count is a key aspect to understand the need for improving the transport facilities for a particular road network. By observing the traffic flow data in the past years and also predicting the future traffic flow from the data, we can build the required transportation model.

Travel is mainly used for transporting goods and people. Number of people traveling at a particular time also plays a major role. We need to find out the average rate of people traveling. The average occupancy then can be used for many calculations. The total number of persons traveling then can be calculated using the number of vehicles and average occupancy of each vehicle. The loss in terms of hours due to the traffic can be found out. Transportation is mostly for movement of goods and passengers. The modern civilization owes a great deal to ease of transportation and communication. For a country to move forward, a good transportation system is very important. Transportation plays a major contributor to the economic, industrial, social, cultural development of the country. The demand for transport is increasing day by day due to the rise in population and economic progress.

The development of a place is dependent upon the physical, social, and institutional infrastructure. Indian cities currently seem to consist mostly of personal transport. But we need to reduce personal transport and instead give importance to public transport. This means we need to increase both quantity as well as quality of public transport and its effective use. Rise in population in Indian cities has led to substantial increase in transport demand. This in turn has caused a high level of pollution in overloaded streets. But the public transport haven't changed much in the last few decades in terms of numbers or management. This has only made people dislike public transport and instead choose their personal transport. Until and unless we make public transport safer and efficient, we will not be able to make a major improvement in the transport domain. (Singh 2005).

## 13.3 Passenger Transport Services

Means of passenger transport may be categorized as (i) City bus service (ii) Intermediate transport services.

**City Bus Services**: These days the city bus transportation has become most used in the majority of all cities, town and semi-urban areas. Mini-buses are also used for brief distances within the cities to alleviate congestion throughout peak hours. Economical and low-cost service may be a necessity for the poor and middle-class families. Correct rules and regulations for smooth functioning of city buses must be provided. It includes (i) correct parking system for the buses in operation within the city or alternative routes. (ii) Regulation in fare rates by the government/authority from time to time. (iii) Proper maintenance of routes to make it convenient for the bus operators to perform their duties well. (iv) Recognition to the bus transport operators and improving their service conditions, safeness and necessary infrastructure, etc. Bus-priority system has inconvenience if the roads are not built properly and consumers take due benefits with intermediate transport services if the price issue is identical and take longer.

**Intermediate Transport Services**: It is the system opposite to the bus priority system where bus services do not seem to be considered favorable. The utilization of auto-rickshaws for brief distance transportation of passengers and goods is on the rise in all cities and towns of many states. Every kind of intermediate transport services are able to operate in the city. Proper rules provided should include the following (i) Offer proper parking for any sort of intermediate transport services within the city. (ii) Offer recognition to the organizations/associations of such operators etc. In spite of rise in auto ownership in the country, the role of bus transportation in moving the people in Indian cities cannot be opposed. Buses do not need any expensive infrastructure and can begin the services the day they are acquired. City bus operations anywhere in the world are financially not usable. Economists showed how returns on investment in public transport ought to be calculated to take into account not solely of the balance of operating revenue and expenditure but also the indirect benefits made by an efficient public transport system. It is these indirect benefits to the community as a whole, and not only for passengers, that justify public authority intervention in public transport finance. Other than this, the ecologists have shown another justification for public contribution towards public transport considering its importance in resolving the present inconveniences to the society of uncontrolled personal traffic ensuing into unendurable pollution and use of already scarce urban space. In reality, some of the western nations like Switzerland, Germany, and Italy consider public transport as a valuable resource in nature conservation and every country concerned has developed its own individual solutions.

Sadly India has not developed any future specific scheme of bus grant for the city bus operations, with the result several of the city buses are in a very poor mechanical condition with small amount of money either for replacement of fleet or addition to the present fleet. In reality, rather than giving grant, the authorities of a number of the Indian city services like Delhi and Calcutta have taken the straightforward route of choosing for personal vehicles rather than subsidizing public transport corporations. Private bus operators, particularly small fleet owners, cause chaotic conditions on the already congested roads. Therefore, the best alternative still remains a public transport efficiently operated with support and grant from the state. Finance for grant may come from increased road tax for cars and two-wheelers. The present tax rate for personalized vehicles in India is just too low for using the expensive services like street lighting, traffic lights, etc. In brief, there is no short way to overcome the current chaotic condition is India aside from increased road tax, levy of parking charges, etc. on motorists. The amount so collected, however, ought to be completely spent for upgrading the public transport systems.

## 13.4 Methodology

(a) **Concepts of fuzzy numbers**

If $x$ is a universe of discourse and $x$ be any particular element of $X$, then a fuzzy set $\tilde{A}$ in $X$ is a set of ordered pairs $\tilde{A} = \{x, \mu_{\tilde{A}}(x)\}, x \in X$ where $\mu_{\tilde{A}}(x)$ is termed membership function of $x$ in $\tilde{A}$. We shall assume that $\mu_{\tilde{A}}(x) : X \to [0, 1]$ with value 1 and 0 representing respectively full membership and non-membership in the fuzzy set $\tilde{A}$. The height h($\tilde{A}$) of a fuzzy set $\tilde{A}$ is the largest membership grade obtained by any element in the set. A fuzzy set $\tilde{A}$ of a set $X$ is said to be normal fuzzy set iff $\mu_{\tilde{A}}(x) = 1$ for at least one $x \in X$. A fuzzy set $\tilde{A}$ of a set $X$ is convex if $\mu_{\tilde{A}}(x)[\beta x_1 + (1-\beta)x_2] \geq \min\{\mu_{\tilde{A}}(x_1), \mu_{\tilde{A}}(x_2)\}, \ \forall\, x_1, x_2 \in X \ and \ \beta \in [0, 1]$.

A fuzzy set is called a fuzzy number if it Satisfied the two properties:

(i) $\mu_{\tilde{A}}(x) = 1$, for at least one $x \in$ R
(ii) Every ordinary subset $\tilde{A}_\alpha = \{x, \mu_{\tilde{A}}(x) \geq \alpha\}, \alpha \in [0, 1]$ is convex.

(b) **Triangular fuzzy numbers**.

A fuzzy number $\tilde{A}$ is a triangular fuzzy number (TFN) which is denoted by $(a_1, a_2, a_3)$, $(a_1 \leq a_2 \leq a_3)$ if its membership function $\mu_{\tilde{A}}$ is given by;

$$\mu_{\tilde{A}}(x) = \begin{cases} 0 & x < a_1 \\ \frac{x-a_1}{a_2-a_1} & a_1 \leq x \leq a_2 \\ \frac{a_3-x}{a_3-a_2} & a_2 \leq x \leq a_3 \\ 0 & x < a_3 \end{cases}$$

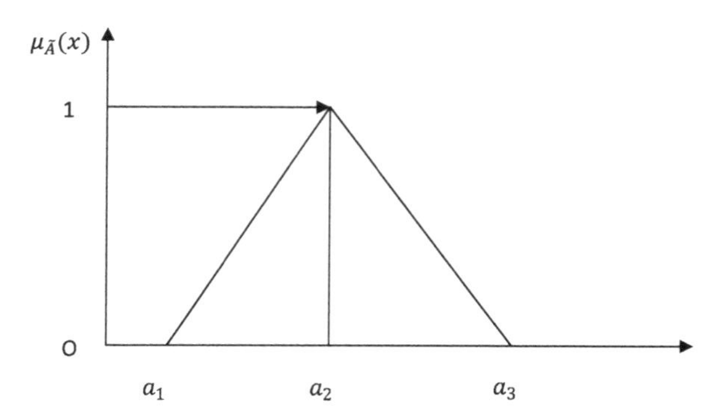


Each fuzzy number can be represented by its $\alpha - cuts$. $\alpha - cuts$ of each fuzzy number are closed intervals of real numbers for all real $\alpha \in [0, 1]$. A TFN can also be represented by $\alpha - cuts$.

Thus for all $\alpha \in [0, 1]$.

$$A_\alpha = [(a_2 - a_1)\alpha + a_1, -(a_3 - a_2)\alpha + a_3]$$

Ordinary number (Defuzzyfication) is given by $\tilde{A} = \frac{a_1+2a_2+a_3}{4}$.

Arithmetic operations on triangular fuzzy numbers are summarized as follows. Suppose $\tilde{A} = (a_1, a_2, a_3)$ and $\tilde{B} = (b_1, b_2, b_3)$ are two given triangular fuzzy numbers. Then.

(1) Sum of two TFN: The addition of $\tilde{A}$ *and* $\tilde{B}$ is

$$\tilde{A}(+)\tilde{B} = \left(a_{1,}a_{2,}a_3\right)(+)\left(b_{1,}b_{2,}b_3\right) = \left(a_1 + b_1, a_2 + b_{2,}a_3 + b_3\right)$$

where $a_1, a_2, a_3, b_1, b_2, b_3$, are any real numbers.

(2) Difference of two TFN: The Subtraction of $\tilde{A}$ *and* $\tilde{B}$ is

$$\tilde{A}(-)\ \tilde{B} = (a_1, a_2, a_3)(-)(b_1, b_2, b_3) = (a_1 - b_3, a_2 - b_2, a_3 - b_1)$$

where $a_{1,}a_{2,}a_{3,}b_{1,}b_{2,}b_3$, are any real numbers.

(3) Multiplication of two TFN: Multiplication of $\tilde{A}$ *and* $\tilde{B}$ is

$$\tilde{A}(\times)\ \tilde{B} = (a_1, a_2, a_3)(\times)(b_1, b_2, b_3) = (a_1b_1, a_2b_2, a_3b_3)$$

where $a_{1,}a_{2,}a_{3,}b_{1,}b_{2,}b_3$, are any non-zeropositive real numbers.

(4) Division of two TFN: Division of $\tilde{A}$ *and* $\tilde{B}$ is
$\frac{\tilde{A}}{\tilde{B}} = (\frac{a_1}{b_3}, \frac{a_2}{b_2}, \frac{a_3}{b_1})$. Where $a_1, a_2, a_3, b_1, b_2, b_3$, are any non-zero positive real numbers.

For multiplication, division, maximum and minimum operations of TFNs cannot be used The computation can be done using confidence interval at each level $\alpha$.

$$\text{Maximum operation of TFNs} \Rightarrow \tilde{A}(\cup)B = (a_1 \cup b_1, a_2 \cup b_2, a_3 \cup b_3)$$
$$\text{Minimum operation of TFNs} \Rightarrow A(\cap)B = (a_1 \cap b_1, a_2 \cap b_2, a_3 \cap b_3)$$

Let $\widetilde{G_1}, \widetilde{G_2}, \widetilde{G_3} \ldots, \widetilde{G_n}$ be the n objectives and $\widetilde{C_1}, \widetilde{C_2}, \widetilde{C_3} \ldots, \widetilde{C_m}$ be m constraints. Then the resultant decision is the intersection of the given objectives $\widetilde{G_1}, \widetilde{G_2}, \widetilde{G_3} \ldots, \widetilde{G_n}$ and the given constraints $\widetilde{C_1}, \widetilde{C_2}, \widetilde{C_3} \ldots, \widetilde{C_m}$. Then $\tilde{D} = \widetilde{G_1} \cap \widetilde{G_2} \cap \widetilde{G_3} \cap \ldots \cap \widetilde{G_n} \cap \widetilde{C_1} \cap \widetilde{C_2} \ldots \cap \widetilde{C_m}$ and correspondingly $\mu_D = \min\{\mu_{G_1}, \mu_{G_2}, \mu_{G_3}, \ldots \mu_{G_n}, \mu_{C_1}, \mu_{C_2}, \ldots \mu_{C_m}\}$.

## 13.5 Analysis with TFNs

When there is a set of finite alternatives $X_1, X_2, X_3, \ldots\ldots\ldots, X_n$ and a set of objectives, $G_1, G_2, G_3, \ldots\ldots\ldots G_m$ having varying degrees of importance then the problem is to select the best alternative. However, the selector gives his decision as to how well the alternatives satisfy the objectives in linguistic terms. The steps of the procedure are briefly outlined as follows (Kumar and Karmaker 2003).

(a) Express the Fuzzy terms/linguistic terms given by the decision maker as triangular fuzzy numbers say as $\tilde{R}_{t_1}, \tilde{R}_{t_2}, \ldots\ldots\ldots\ldots \tilde{R}_{t_n}$

(b) Express most important objective, important objective, less important, not all important, etc. as triangular fuzzy numbers as $\tilde{t}_1, \tilde{t}_2, \ldots\ldots..\tilde{t}_m$

(c) Let us find relative weights by finding the ratio of each triangular fuzzy number with the composite triangular number i.e. if

$$\tilde{t}_1 = (p_1, p_2, p_3), \quad \tilde{t}_2 = (q_1, q_2, q_3), \quad \ldots\ldots\ldots\ldots\ldots\ldots \tilde{t}_m = (r_1, r_2, r_3)$$

are m triangular fuzzy number(TFN), then the composite TFN is given by

$$\tilde{T} = \tilde{t}_1 + \tilde{t}_2 + \ldots + \tilde{t}_m$$
$$\tilde{T} = (p_1 + q_1 + \ldots + r_1, p_2 + q_2 + \ldots + r_2, \; p_3 + q_3 + \ldots + r_3)$$

Then relative weights are given by

$$w_1 = \tilde{t}_1(:)\tilde{T}$$
$$w_2 = \tilde{t}_2(:)\tilde{T}$$
$$\ldots\ldots\ldots\ldots\ldots$$
$$w_m = \tilde{t}_m(:)\tilde{T}$$

Operation of division on TFNs does not necessarily give a TFN. So, we can approximate the results as

$$W_1^* = \left(\frac{p_1}{p_1 + q_1 + \ldots\ldots + r_1}, \frac{p_2}{p_2 + q_2 + \ldots\ldots + r_2}, \frac{p_3}{p_3 + q_3 + \ldots\ldots + r_3}\right)$$

Similarly, we find $W_2^*, W_3^*, \ldots\ldots\ldots W_m^*$.

(d) Find the normal ordinary number corresponding to each relative weight say if $\tilde{A} = (a_1, a_2, a_3)$ is a triangular fuzzy number, then its normal ordinary number is given by

$$\hat{A} = \frac{a_1 + 2a_2 + a_3}{4}.$$

Let $\tilde{W}_1, \tilde{W}_2, \ldots\ldots\ldots\ldots\ldots, \tilde{W}_m$ be the corresponding normal ordinary number.

(e) Let us calculate

$$G_1^{\hat{W}_1} = \{\tilde{R}_{1_1}^{\hat{W}_1}, \tilde{R}_{2_1}^{\hat{W}_1}, \ldots\ldots\ldots\ldots\ldots \tilde{R}_{n_1}^{\hat{W}_1}\}$$
$$G_2^{\hat{W}_2} = \{\tilde{R}_{1_2}^{\hat{W}_2}, \tilde{R}_{2_2}^{\hat{W}_2}, \ldots\ldots\ldots\ldots\ldots \tilde{R}_{n_2}^{\hat{W}_2}\}$$
$$\ldots\ldots\ldots\ldots\ldots\ldots\ldots\ldots\ldots\ldots\ldots\ldots\ldots\ldots\ldots\ldots$$
$$G_m^{\hat{W}_m} = \{\tilde{R}_{1_m}^{\hat{W}_m}, \tilde{R}_{2_m}^{\hat{W}_m}, \ldots\ldots\ldots\ldots\ldots \tilde{R}_{n_m}^{\hat{W}_m}\}$$

Then resultant decision model is.

$$\mathrm{D} = G_1^{\hat{W}_1} \cap G_2^{\hat{W}_2} \cap \ldots\ldots\ldots \cap G_m^{\hat{W}_m}.$$

Where decision D is a fuzzy subset of potential systems.

(f) Let us find the normal ordinary numbers corresponding to the Triangular Fuzzy numbers thus obtained and Carry out linear ordering of the normal ordinary numbers and choose system $X_i$ having highest normal ordinary number.

## 13.6 Applications

Under different routing and scheduling systems.

Let $X_1 and \quad X_2$ be a set of alternatives or systems where $X_1$ represents Bus priority system in the city and $X_2$ represents non-bus priority system. Let there be five objectives $G_1$, $G_2$, $G_3$, $G_4$ and $G_5$ to be satisfied by each alternative where:

$G_1$ Given mode should have a cost which is feasible to the commuters.
$G_2$ Given mode should have time minimization.
$G_3$ Given mode should be less pollution in the environment.
$G_4$ Given mode should be feasible to the commuters in the summer and winter season.
$G_5$ Given mode should be good looking and comfortable.

- Further, it is given that $G_1$ is the most important objective, $G_2$ is a very important objective, $G_3$ is important objective, $G_4$ is the less important objective, and $G_5$ is not so important objective. There is a single decision-maker who has to decide which alternative is to be purchased. The decision-makers find it easier to express his viewpoint about these alternatives, as to how best they satisfy the objectives in linguistic terms as $X_1$ satisfies: $G_1$ very well, $G_2$ poorly, $G_3$ very well, $G_4$ fairly well and $G_5$ Extremely well. $X_2$ satisfies: $G_1$ fairly well, $G_2$ Extremely well, $G_3$ very poorly.$G_4$ poorly, $G_5$ fairly well

We now express these linguistic terms and the linguistic terms associated with the importance of the objectives as triangular fuzzy numbers [Tables 13.1 and 13.2] (Zadeh 1976) (Figs. 13.1 and 13.2).

Adding $\tilde{t}_1, \tilde{t}_2, \tilde{t}_3, \tilde{t}_4, \tilde{t}_5$ we get (2.0, 2.7, 3.2) as the cumulative triangular fuzzy number. Then approximate relative weights are calculated as:

**Table 13.1** The corresponding fuzzy numbers for the Linguistic terms

| Linguistic terms/fuzzy terms | (TFN) |
|---|---|
| More poorly | (0, 0,0.3) |
| Poorly | (0, 0.3, 0.5) |
| Well | (0.2, 0.5, 0.8) |
| Very well | (0.5, 0.7, 1) |
| Extremely well | (0.7, 1, 1) |

**Table 13.2** The corresponding fuzzy numbers for the objectives

| Linguistic terms/fuzzy terms | Triangular fuzzy numbers(TFN) |
|---|---|
| Most important | $(0.7, 0.9, 1) = \tilde{t}_1$ |
| Very important | $(0.6, 0.8, 0.9) = \tilde{t}_2$ |
| Important | $(0.4, 0.5, 0.6) = \tilde{t}_3$ |
| Less important | $(0.2, 0.3, 0.4) = \tilde{t}_4$ |
| Not so important | $(0.1, 0.2, 0.3) = \tilde{t}_5$ |

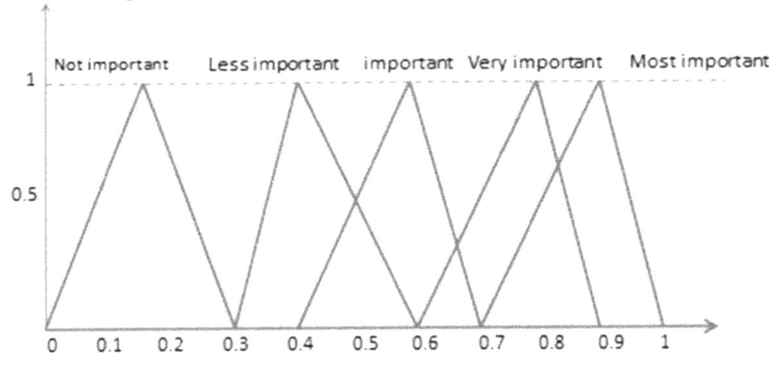


**Fig. 13.1** Representation of Triangular fuzzy numbers for different objectives of varying degrees

$$\tilde{w}_1 = \left(\frac{0.7}{3.2}, \frac{0.9}{2.7}, \frac{1}{2.0}\right) = (0.21875, 0.33333, 0.5)$$

$$\tilde{w}_2 = (0.1875, 0.2962, \ 0.45), \tilde{w}_3 = \ (0.125, \ 0.1851, \ 0.3),$$

$$\tilde{w}_4 = (0.0625, \ 0.111, \ 0.2)$$

$$\tilde{w}_5 = (0.03125, \ 0.0740, \ 0.15)$$

Associated ordinary numbers corresponding to these relative weights are given as

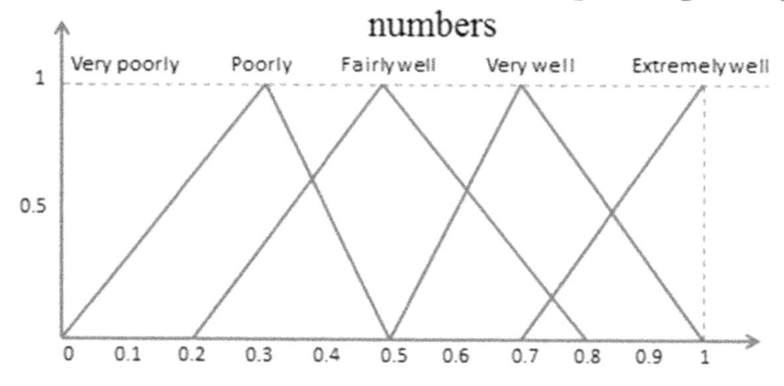


**Fig. 13.2** Representation of Triangular fuzzy numbers of linguistic terms satisfied by different alternatives

$$w_1 = \frac{0.21875 + 2 \times 0.3333 + 0.5}{4} = 0.346175,$$
$$w_2 = 0.307475, \; w_3 = 0.1988, \quad w_4 = 0.121125,$$
$$w_5 = 0.0823125, \; \sum_{i=1}^{3} w_i = 1$$

Next, let us find

$$\begin{aligned}
G_1^{w_1} &= \left\{(0.5.0.7, 1)^{0.346}(0, 0.3, 0.5)^{0.346}\right\} \\
&= \{(0.786, 0.883, 1)(0, 0.659, 0.786)\} \\
G_2^{w_2} &= \left\{(0, 0.3, 0.5)^{0.307}(0.7, 1, 1)^{0.307}\right\} \\
&= \{(0, 0.690, 0.808)(0.896, 1, 1)\} \\
G_3^{w_3} &= \left\{(0.5, 0.7, 1)^{0.198}(0, 0, 0.3)^{0.198}\right\} \\
&= \{(0.871, 0.931, 1)(0, 0, 0.787)\} \\
G_4^{w_4} &= \{(0.2, 0.5, 0.8)^{0.121}(0, 0.3, 0.5)^{0.121} \\
&= \{(0.832, 0.919, 0.973)(0, 0.864, 0.919)\} \\
G_5^{w_5} &= \left\{(0.7, 1, 1)^{0.082}(0.2, 0.5, 0.8)^{0.082}\right\} \\
&= \{(0.971, 1, 1)(0.876, 0.944, 0.981)\}
\end{aligned}$$

Now

$$G = G_1^{w_1} \cap G_2^{w_2} \cap G_3^{w_3} \cap G_4^{w_4} \cap G_5^{w_5} = \{(0, 0.690, 0.808)(0, 0, 0.786)\}$$

$$Associated\ ordinary\ number(Bus\ priority\ system) = \frac{0 + 2 \times 0.690 + 0.808}{4} = 0.547$$

$$Associated\ ordinary\ number(Non\text{ - }Bus\ priority\ system) = \frac{0 + 2 \times 0 + 0.786}{4} = 0.196$$

## 13.7 Conclusion

The associated ordinary numbers corresponding to these triangular fuzzy numbers are given as D = (0.547, 0.196).Since the associated ordinary number corresponding to $X_1$ is the highest, system $X_1$ is superior to system $X_2$ i.e. Towns/cities providing bus priority system for the passengers traveling from place to place within the city have better than intermediate transport system (auto/taxi/car etc.). By considering the factors of traveling cost, less pollution affect, travel time it can be concluded that the commuters have advantages in Bus priority system than that of non- Bus priority system. To ensure Bus priority system, the following steps should be taken up a) improvement of roads, widening of roads should be taken up as top priority b) proper Bus parking should be provided outside the main business area and c) Permitted to run with limited stopping time.

## References


Brindhavanam M, Rosario GM (2017) A fuzzy comparison method for an deterministic inventory model without shortages. Int J Comput Appl Math 12(1)

Dubois D, Prade H (1978) Operations of fuzzy number's. Int J Syst Sci 9(6):613–626

Dubois D, Prade H (1980) Fuzzy sets and systems. Theory and applications. Academic press, New York

Feng D, Nam JX, Zhang M (Oct 2010) A ranking method of triangular intuitionistic Fuzzy numbers and application to decision making. Int J Comput Intell Syst 3(5):522–530

Gil-Lafuente AM, Merigó JM (2010) Decision making techniques in political management. In: Fuzzy optimization. Studies fuzziness soft computer, vol 254. Springer, Berlin, pp 389–405

Kaufmann A (1975) Introduction to theory of Fuzzy subsets, vol 1. Academic press, New York

Kaufmann A, Gupta MM (1985) Introduction to Fuzzy arithmetic. Van Nostrand Reinhold, New York

Kuchta D (2010) The use of fuzzy numbers in practical project planning and control in Fuzzy optimization. Studies fuzziness soft computer, vol 254. Springer, Berlin, pp 323–339

Kumar A, Karmaker G (2003) Optimal system selection using Fuzzy numbers. Int J Manage Syst 19(3)

Nguyen HT (1978) A note on extension principle for fuzzy sets. J Math Anal Appl 64:369–380

Prangishvili A, Tsabadze T, Tsamalashvili T (2015) Application of fuzzy sets in solving some problem of management, I. J Math Sci 208(6):661–676

Singh SK (2005) Review of urban transportation in India. J Public Transp 8(1)
Zadeh LA (1965) Fuzzy sets. Inf Control 8:339–353
Zadeh LA (1975/1976) The concept of a Linguistic variable and its applications to approximate reasoning—parts I,II and III. Inform Sci 8:199–249; 301–357; 9:43–80
Zadeh LA (1978) Fuzzy sets as a basis for a theory of possibility. Fuzzy Sets Syst 1:3–28
Zadeh LA, Bellman RE (1970) Decision making in a fuzzy Environment. Manage Sci 17:140–164
Zhang F, Joshua, Lim CP, Zhao Y (2014) A new method for ranking fuzzy numbers and its application to group decision ranking. Appl Math Model 38:1563–1582
Zimmerman HJ (2001) Fuzzy set theory and its applications. Springer Science Business Media, LLC

Lightning Source UK Ltd.
Milton Keynes UK
UKHW050010310822
408099UK00006B/50

9 789811 617317